I0408533

THE FOOD OF THE FUTURE

VERTICAL FARMING

&

LAB-GROWN MEAT

DAVID SANDUA

The food of the future: Vertical farming & lab-grown meat
© David Sandua 2023 All rights reserved
eBook & Paperback Edition

"The future of food is the food of the future."

Norman Borlaug.

INDEX

I. INTRODUCTION

Food is an essential part of human existence, providing nourishment to our bodies and fuel to our minds. The way we produce and consume food has a significant impact on our health, the environment, and the economy. With the world population expected to reach 9.7 billion by 2050, the demand for food is increasing rapidly. Our current food production system is unsustainable and poses challenges such as land degradation, water shortages, and greenhouse gas emissions. To address these challenges and meet the growing demand for food, innovations in food production are crucial. Two of the most promising innovations are lab-grown meat and vertical farming. Lab-grown meat is produced by culturing animal cells in a lab, while vertical farming is a method of growing crops in vertically stacked layers using controlled environments. These innovations have the potential to revolutionize the way we eat, offering a sustainable and efficient solution to food production. In this essay, we will explore the potential benefits and drawbacks of these two innovations and how they may change the way we eat.

BACKGROUND INFORMATION ABOUT THE ISSUE OF FOOD PRODUCTION AND CONSUMPTION

The global food system has faced numerous challenges that have shaped the way we think about how we produce and consume food. One of the biggest problems we face is the issue of production and consumption of traditional meat products. The relationship between meat production and consumption plays an important role in climate change, greenhouse gas emissions, and animal welfare concerns. The livestock industry continues to generate high levels of greenhouse gases, which has been responsible for a significant portion of environmental degradation around the world. Animal agriculture is responsible for roughly one-fifth of global greenhouse gases, a statistic that prompts concern for the future of the planet. This relationship has led to a push to explore alternative methods of food production and consumption that can address these pressing concerns, while providing a sustainable and ethical solution. The debate around the future of food and how we produce and consume it becomes a crucial issue in the 21st century and one that requires a multidisciplinary approach. Consequently, when exploring food production and consumption trends, it is essential to consider the limitations of current practices and how innovations in the food industry could revolutionize the way we eat. Vertical farming is a burgeoning trend in food production that has captured the attention of policymakers, entrepreneurs and consumers alike. As the name suggests, vertical farming involves growing crops on vertical surfaces, thus making the most effective use of space available. Vertical farming allows for greater food production densities and crop yields than traditional farming methods. This approach to food production is also more resource-efficient, with significantly less land, water and fertilizers needed to grow the same amount of crops. The vertical farming system can be strategically placed in urban centers where distance from a farm to the point of consumption is minimized. This provides for the potential for large reductions in the carbon footprint of food production, as well as the possibility of increased access to fresh produce in farmer's market settings. Vertical farming is a trend that is gaining steady momentum, and as technology continues to advance and costs fall, it may well become a significant part of our food systems in the coming years. Lab-grown meat, also known as clean meat or cultured meat, is an innovation that could revolutionize the meat production industry. Lab-grown meat is meat produced by using stem cells isolated from animals, which are then grown in a laboratory under controlled conditions. The result is a product that looks and tastes like traditional meat but that has been grown without raising animals for slaughter. The process can create high-quality meat products with far fewer emissions than traditional meat products. Clean meat is a promising technology that could significantly reduce the ecological footprint of the agriculture industry. It is not without its challenges, including cost, scaling up production, and regulatory agencies. Lab-grown meat, while still in its early stages of production, is a technology that has captured significant attention from investors and consumers both in the United States and internationally. As it continues to gain traction and find more widespread acceptance, it has the potential to revolutionize the way we produce and consume meat in the coming years.

Food production and consumption have long been debates on a planet where millions go hungry every day while others are morbidly obese. Around one-third of all food produced globally is either lost or wasted after production, due to supply chain inefficiencies, food safety issues, and plain personal habits. Similarly, as world incomes rise, so too does meat consumption, resulting in both health and agricultural sustainability concerns. In the absence of change, it is likely that these issues will only grow more severe. The future of food production and consumption will depend on our collective ability to address this challenge, and the innovations in food production brought by emerging technologies are part of the solution. New ideas, such as vertical farming and lab-grown meat, are areas of immense interest where we may well be able to take significant steps towards a sustainable and ethical food system. The problem with our current food system is that traditional meat production yields unsustainable results, and the increasing demand for meat across the world is incredibly problematic. The intensive farming practices associated with livestock farming lead to the development of antibiotic resistance, zoonotic diseases, and environmental pollution. Land and water usage constitute significant inputs for agriculture, putting future farming resources at risk. The cost of meat production is increasing, creating pressure on farmers to set competitive prices and forcing them to choose low-cost methods, such as factory farming, which actively contribute to the environmental concerns noted above. In response to these challenges, vertical farming and lab-grown meat represent two potential, innovative solutions to change the way we eat and produce food. The challenges associated with food production –

such as resource overuse, animal welfare, and environmental degradation – can have a substantial impact on their viability and individual consumption decisions. Increasing awareness of alternative practices and more open dialogue about risks and benefits will be essential in navigating these challenges. Innovation in food production and consumption serves one crucial goal – meeting the most basic human needs in a way that doesn't destroy nature.

DEFINITION OF LAB-GROWN MEAT AND VERTICAL FARMING

Lab-grown meat, also known as cultured meat or cell-based meat, refers to meat that is produced through the cultivation of muscle tissue cells in a laboratory setting. The process involves taking a small sample of muscle tissue from an animal and then culturing these cells in a medium that allows them to grow and differentiate into muscle fibers. This allows the production of meat without the need for the whole animal to be raised, slaughtered, and processed. By doing so, lab-grown meat provides a more ethical and sustainable solution to meat production, as it eliminates the need for animal breeding and slaughter. It has the potential to reduce the environmental impact of meat production by decreasing greenhouse gas emissions, land use, and water consumption. Lab-grown meat is also expected to offer significant nutritional benefits, as it can be customized to have specific fat and nutrient profiles. Several companies have emerged that are actively working on developing lab-grown meat products. Among these are Memphis Meats, Mosa Meat, Future Meat Technologies, Aleph Farms, and SuperMeat. These companies have already produced prototypes of various types of lab-grown meat, including beef, chicken, and pork. Due to the high production costs involved, lab-grown meat is currently much more expensive than conventional meat. Its taste, texture, and appearance are still being refined to match that of natural meat. Despite these challenges, many experts believe that lab-grown meat has the potential to revolutionize the food industry and transform the way we eat meat. Vertical farming, on the other hand, is a method of growing crops in vertically stacked layers within a controlled environment, such as a high-rise building, a shipping container, or a greenhouse. This approach allows for year-round crop production, regardless of weather conditions, and can significantly increase crop yields per unit of land area. Vertical farming can also reduce water consumption by up to 70%, as the closed-loop system recycles water and nutrients and reduce the use of pesticides and herbicides by eliminating exposure to environmental factors such as wind and soil erosion. Because crops are grown indoors, vertical farms are not subject to the unpredictable weather conditions that can lead to crop failures and food shortages. One of the most prominent companies in the vertical farming industry is AeroFarms, a New Jersey-based company that has developed a highly efficient vertical farming system that uses only 5% of the water required for conventional farming and yields up to 390 times more crops per square foot. AeroFarms uses a technology called "aeroponics" that relies on misting plants' roots with a nutrient-rich solution. This system allows for precise control over the amount of water, light, and nutrients that plants receive, resulting in faster growth rates and fewer plant diseases. Other vertical farming companies, such as Plenty and BrightFarms, are also gaining traction, and several startups are focused on developing automated vertical farming systems that use artificial intelligence and robotics to manage crop growth. Together, lab-grown meat and vertical farming represent two of the most promising innovations in food production, with the potential to disrupt traditional agriculture and radically transform the way we produce and consume food. While they both face significant challenges in terms of cost, scalability, and consumer acceptance, they offer a glimpse into a more sustainable and efficient future of food. By providing a more ethical and sustainable alternative to traditional meat production and increasing crop yields while reducing environmental impact, these technologies offer hope for creating a more secure and abundant food supply while reducing the industry's carbon footprint. Lab-grown meat and vertical farming could play a critical role in addressing some of the most pressing challenges facing our food system in the coming decades. For instance, as the global population continues to grow rapidly, and the demand for food increases, the conventional agricultural practices will be strained to keep up. By contrast, the scalability of vertical farming enables it to produce large quantities of food in a small space. Similarly, lab-grown meat has the potential to feed the world's growing population without expanding the lands dedicated to animal agriculture, which currently accounts for more than 40% of global land use. Lab-grown meat and vertical farming are two innovations that show incredible potential to transform modern agriculture and create a more sustainable, secure, and efficient future of food. They both address critical problems related to environmental impact, animal welfare, food safety, and production efficiency. While still in the early stages of development, these technologies hold the promise of a more equitable food system that balances human needs with environmental stewardship and animal welfare. As we look towards the future of food, we should not overlook the critical role of lab-grown meat and vertical farming in promoting a sustainable and just food system for all.

THE PURPOSE OF THE ESSAY

In summary, the purpose of this essay is to evaluate the latest advancements in the food industry, such as lab-grown meat and vertical farming, and examine how they will affect the future of food production and consumption. Through an analysis of peer-reviewed literature, news articles, and expert opinions, we aim to provide a comprehensive overview of this emerging field and the potential implications it has for sustainable agriculture, animal welfare, and global health. Firstly, the essay will cover lab-grown meat, also known as cultured meat, and discuss how it differs from traditional animal agriculture, its advantages and challenges, and its potential impact on animal welfare, the environment, and public health. We will also explore the role of biotechnology in culturing meat and the various methods used to produce it, as well as the regulations and public perceptions surrounding its commercialization. Secondly, the essay will explore vertical farming, an innovative method of growing crops in vertically stacked layers using artificial lighting, hydroponics, and other advanced technologies. We will analyze the benefits and limitations of this farming approach, including its potential for increased crop yields, reduced water usage, and decreased transportation costs, as well as its challenges in terms of energy use and initial investments. We will examine how these two innovations are expected to impact the wider food ecosystem, from the way we produce and consume food to the implications for the food industry, farmers, and consumers. The essay will also explore the potential commercialization and adoption of these technologies, and their impact on global food security and sustainability. This essay seeks to provide an informed and critical analysis of the future of food, and the potential benefits and limitations of these emerging innovations. By doing so, we hope to contribute to the ongoing conversation on sustainable food production and the role of technology in shaping our food system. Vertical farming and lab-grown meat are two exciting innovations that could drastically change the way we produce and consume food. Vertical farming, the practice of growing crops or raising animals in vertically stacked layers, has the potential to reduce farmland use, increase urban agriculture, and increase food production efficiency. Lab-grown meat, which is produced by growing animal muscle tissue in a laboratory setting, offers a solution to some of the environmental and ethical concerns associated with traditional animal agriculture. While these technologies are not without their challenges, they offer exciting possibilities for the future of food. Vertical farming offers many potential benefits, particularly in urban areas where space for farmland is limited. By growing crops and raising animals in vertical layers, it's possible to increase the amount of food that can be produced in a given area. Vertical farms use significantly less water than traditional farming methods and rely on hydroponic or aeroponic systems to provide the necessary nutrients to crops. This means that they can be much more efficient in their use of resources than traditional farming. Another benefit of vertical farming is the increased control it offers over growing conditions, including temperature, humidity, and lighting. This can lead to higher crop yields and more consistent quality produce. Vertical farming can be used to grow crops year-round, allowing for more consistent and reliable food production. Despite these benefits, challenges exist in implementing vertical farming on a large scale. The initial cost to set up these farms can be high and may be prohibitive for some farmers or communities. The energy requirements to operate these farms can also be a concern, given that lighting and climate control are necessary for optimal growth. Nonetheless, with emerging technologies in renewable energy and efficient LED lights, the energy requirements of these farms are gradually reducing. There are also some limitations to the type of plants that can be grown in a vertical farm. While vegetables and herbs are great candidates for vertical farming, things like grains and certain fruits are not as easily grown in this type of system. Lab-grown meat is another innovation in food production that offers a promising alternative to traditional animal agriculture. Growing meat in a lab setting involves taking stem cells from live animals and then using these cells to grow muscle tissue in a culture dish. This tissue can then be used to create meat products, without the need for animal slaughter. This technology offers several potential benefits. For one, it significantly reduces the environmental impact of meat production. Meat farming can be an incredibly resource-intensive process, using large amounts of water, feed, and energy to raise animals. Animal agriculture is responsible for a significant portion of greenhouse gas emissions worldwide. Lab-grown meat production, meanwhile, uses significantly less water and energy. Lab-grown meat also offers a solution to some of the ethical concerns associated with traditional meat production. Animal welfare has been an ongoing issue of concern to many groups for years, including animal rights activists. This method of meat production could help to reduce animal suffering. It also enables the creation of more diverse meat products, potentially allowing for the growth of exotic or endangered species without the need for harmful animal farming practices. Despite the promise of lab-grown meat, several challenges persist. One of the biggest obstacles is the cost of producing meat in this way. Currently, lab-grown meat is significantly more expensive than traditional meat, making it difficult to bring to market at a competitive price. There are concerns around the safety of the lab-grown meat, particularly regarding how it is produced. As the technology is still relatively new, it is still unclear whether current production methods could lead to contamination or other safety concerns. The innovations of vertical farming and lab-grown meat offer exciting possibilities for the future of food production and consumption. Both technologies offer solutions to some of the challenges of traditional food production, including land and resource use, environmental impact, and animal welfare. There are still many challenges to overcome before these innovations can become widely adopted. It will be important to continue research to address these issues and improve the efficiency and safety of both technologies. If we can overcome these obstacles, the future of food could be much more sustainable, ethical, and efficient than ever before.

II. THE CURRENT STATE OF FOOD PRODUCTION

The current state of food production has significant implications for the future of humanity. With increasing population growth and climate change, the agriculture industry is facing several challenges. Land degradation, water scarcity, and decreased soil fertility are just some of the problems that are affecting conventional farming. As such, innovation in the food production sector is essential. One of the most promising innovations in the food industry is lab-grown meat. This revolutionary method of producing meat could significantly reduce the environmental impact of the traditional meat industry. Currently, the traditional meat industry is responsible for up to 15% of global greenhouse gas emissions. Cattle ranching is unsustainable in the long run and is one of the primary drivers of deforestation. Lab-grown meat, however, is produced using tissue engineering techniques. This means that cells are cultured in controlled environments, eliminating the need for large-scale animal farming. By reducing the number of animals required to produce meat, lab-grown meat has the potential to reduce the environmental impact of the meat industry significantly. The technology behind lab-grown meat has been around for some time, but it has only recently become more commercially viable. Several companies, such as Memphis Meats and Mosa Meat, have made significant strides in the development of lab-grown meat. While it may sound like science fiction, the method for producing lab-grown meat is relatively straightforward. First, a small sample of animal cells, such as muscle cells, are extracted from the animal in question. The cells are then grown in a lab using a culture medium that contains essential nutrients for the cells to grow and divide. Over time, these cells eventually form muscle tissue and can be used to make meat products. While lab-grown meat still faces several challenges, such as regulatory and societal issues, its potential to revolutionize the meat industry cannot be overstated. With lab-grown meat, it may be possible to produce meat that is indistinguishable from traditional meat while having a much smaller environmental impact. This, in turn, could lead to a reduction in the size of the meat industry, as fewer animals would be required to produce meat products. As such, lab-grown meat has the potential to reshape not only the meat industry but also the agricultural sector as a whole. Another promising innovation in food production is vertical farming. With land degradation and urbanization being such significant factors in the current state of food production, vertical farming could provide a solution to both these problems. Vertical farming is the practice of growing crops in vertically stacked layers using various technologies such as hydroponics, aeroponics, and aquaponics. This allows for year-round cultivation of crops in controlled environments, regardless of climate or weather conditions. Vertical farming also requires very little water, as the water used in the system is continually recycled, resulting in much greater water efficiency than traditional farming. Like lab-grown meat, vertical farming also has the potential to address several issues facing the agricultural sector. The controlled environment in which crops are grown in vertical farms means that fewer pesticides and herbicides are needed, reducing the amount of harmful chemicals entering the ecosystem. Vertical farming allows for much greater

24

crop yields per square foot of land than traditional farming, potentially providing a more sustainable solution to the problem of land degradation. Vertical farming has the potential to supply fresh produce to urban areas where there is a lack of access to fresh produce. This could address the food security issue and provide fresh produce to areas that have previously relied on imports of fresh fruits and vegetables. Vertical farming is not without its challenges, however. The technology and infrastructure required to set up vertical farms can be costly, making it inaccessible to some farmers. There are questions around the long-term sustainability of vertical farming systems, particularly regarding their energy consumption and the environmental impact of producing the materials required for vertical farming structures. The innovations in food production, such as lab-grown meat and vertical farming, have the potential to reshape the food industry in ways that were previously unimaginable. These methods of food production could address several environmental challenges facing the industry, such as land degradation and greenhouse gas emissions. As technology continues to develop and improve, the challenges and limitations of these innovations in food production will be addressed, paving the way for a more sustainable and environmentally friendly food industry. While there are still challenges to be addressed, it is clear that lab-grown meat and vertical farming offer significant hope for the future of food production.

TRADITIONAL FARMING PRACTICES AND THEIR LIMITATIONS

Traditionally, farming practices have been in use for thousands of years. These practices are highly reliant on soil fertility, weather patterns, water resources, and other natural elements. While this method of farming has sustained humanity for a long time, it is not without its limitations. One major limitation is the reliance on crop rotation, a practice in which different crops are grown on the same plot of land in successive seasons to promote soil health, reduce pests and diseases, and provide diversity in the farmers' output. Crop rotations are not always practical in many agricultural settings. For instance, small farmers often have limited land sizes, and they may be forced to plant the same crops over and over for years, degrading the soil's quality. Such farmers may face difficulties obtaining fertilizer, which can be costly, or they may not have the knowledge or skills to grow crops that require rotation. Another limitation of traditional farming practices is the amount of land required to produce enough food to sustain populations. While farming takes up a significant percentage of the world's land, estimates suggest that the world's population will reach 9.8 billion by 2050. This figure means that more land will be required for farming to produce enough food to feed these populations. The amount of arable land available is limited, and competition for it continues to grow as urbanization and other land uses continue to limit the availability of farmland. Climate change will adversely affect some of the arable land, making it impossible to farm in some regions, further limiting the available land.

The water resource limits are another significant limitation of traditional farming practices. Although rainwater moisture is the primary source of water for agriculture, water scarcity has become a significant issue worldwide due to frequent droughts and population growth. Farmers, particularly those in arid regions, must rely on inefficient irrigation methods such as flood irrigation or traditional surface irrigation that result in a substantial amount of water being lost to evaporation, runoff, or seepage. These practices not only lead to water wastage but also deplete the water table and harm surrounding ecosystems. Chemical fertilizers and pesticides used in traditional farming inadvertently pollute groundwater used for both agricultural and domestic use. This situation has led to food contamination and poses a health risk to the public. Other limitations of traditional farming practices include the reliance on manual labor, which has a limit on how much food can be produced at a time. This limitation creates a considerable labor demand, especially in regions where populations are expanding. Farmers must contend with weather patterns that are often unpredictable, leading to crop failure, and farmers' income is volatility low. Traditional farm practices do not encourage consistent yields, particularly because intensive farming is not suitable for preserving soil health. This reason means that farmers have to work harder and use more resources to plant, insure, and maintain crops than they would with alternative farm productions. Traditional farming practices are becoming increasingly limited and unsustainable while population growth and demands for resources continue to increase. The adoption of innovative food production methods like vertical farming and lab-grown meat promises to change the way we approach food production and potentially mitigate some of the limitations of traditional agriculture. While these methods have yet to be embraced at scale, the potential benefits should not be overlooked. Vertical farming allows for food production in three dimensions without relying on large tracts of land. This method of farming uses hydroponics and aeroponics, making it highly efficient in water and nutrient usage. Lab-grown meat can play a significant role in addressing many of the limitations of traditional animal farming practices. If it becomes practical, lab-grown meat production could reduce land, water, and feed required while using fewer pesticides and chemicals that are harmful to the environment. The adoption of innovations in food production is necessary to address the significant challenges posed by traditional farming practices' limitations. The innovation of vertical farming and lab-grown meat production have the potential to alleviate the constraints of traditional farming practices while offering sustainable alternatives that produce healthy, nutritious, and environmentally friendly food. These innovations will change the way we approach food production, and the effect is anticipated will have a long-lasting impact on humanity's food systems.

THE ENVIRONMENTAL IMPACT OF CURRENT FOOD PRODUCTION PRACTICES

Another critical issue with current food production practices is the impact they have on the environment. Modern agriculture has become one of the biggest contributors to greenhouse gas emissions, land, and water degradation, deforestation, and bio-diversity loss globally. According to the Environmental Protection Agency (EPA), the agriculture sector can account for up to 9% of total U.S. greenhouse gases, primarily from the production of methane and nitrous oxide from livestock manure and fertilizer use. The production, processing, and transportation of food also consume significant amounts of fossil fuels, contributing further to the release of greenhouse gases into the atmosphere.

The use of synthetic fertilizers and pesticides has negative impacts on soil health, water quality, and aquatic ecosystems. According to a report by the Food and Agriculture Organization (FAO) of the United Nations, nutrient pollution from agriculture has led to the creation of more than 400 oceanic "dead zones," areas where oxygen levels are too low to support marine life. Pesticides can also accumulate in soil and water, leading to damage to ecosystems and wildlife, as well as potential human health risks. Another environmental concern is the destruction of habitats for agricultural expansion. According to the World Wildlife Fund (WWF), about half of the world's land surface has been converted to support human populations, mostly for agriculture. In Brazil, for example, the expansion of soybean farming, mainly used for animal feed, has resulted in the destruction of millions of hectares of rainforest. The loss of forest cover contributes to climate change, affects soil health and water cycles, and leads to the loss of biodiversity. The water requirements for current agricultural practices are unsustainable, especially in regions experiencing water scarcity. Irrigation accounts for about 70% of all freshwater use globally, with agriculture accounting for about 80% of that amount, according to the FAO. This overuse of water has led to the depletion of many groundwater reserves, which are essential for human consumption and ecosystem health. Climate change is exacerbating this situation, with more frequent droughts and higher temperatures leading to reduced water availability in many regions. Food waste is also a significant contributor to environmental degradation. Around a third of all food produced globally is wasted, representing a significant waste of resources, including water, land, and energy. Food waste also generates methane emissions when it decomposes in landfills, further contributing to greenhouse gas emissions.

In response to these environmental issues, innovations in food production, such as lab-grown meat and vertical farming, have been proposed as sustainable alternatives. Lab-grown meat is produced by growing muscle tissue in a lab setting instead of in animals' bodies. The process involves taking muscle cells from an animal and feeding them nutrients in a controlled environment. As the cells multiply, they form muscle tissue that can be harvested for consumption. Lab-grown meat has several potential benefits compared to traditional meat production. It requires significantly less land and water and generates fewer greenhouse gas emissions, as it does not rely on animal feed production, transportation, and waste. It also eliminates the need for animal slaughter and can potentially reduce the spread of animal-borne diseases. The production process is not yet cost-effective, and the long-term impacts on human health and the environment are still unclear. Vertical farming is another innovative approach with potential benefits for sustainable food production. It involves growing crops in vertically stacked layers, often in an urban setting, using artificial lighting and controlled environmental conditions. This approach has several advantages over traditional farming. It significantly reduces land use, as crops can be grown in a more compact space, and allows for year-round crop production, independent of seasonal weather conditions. It also eliminates the need for pesticides and herbicides, can reduce water use by up to 70%, and reduces transportation costs. Vertical farming also faces several challenges, including the high cost of production, the need for affordable and sustainable energy sources, and the limited variety of crops that can be successfully grown in a controlled environment. The increased reliance on technology in the production process raises questions about food security and the potential for power outages or system failures to disrupt food supply. Current food production practices have significant environmental impacts, ranging from greenhouse gas emissions to water scarcity, nutrient pollution, and habitat destruction. Innovations in food production, such as lab-grown meat and vertical farming, offer potential solutions to these concerns. Both approaches face technical, economic, and societal challenges that must be addressed to ensure their viability and sustainability. The future of food production will undoubtedly require a multifaceted approach that balances environmental, economic, and social considerations to meet the world's growing demand for food while preserving planetary health.

THE IMPACT OF FOOD PRODUCTION ON HUMAN HEALTH

Food production has a direct impact on human health, and it is important to consider how innovations in food production, such as lab-grown meat and vertical farming, will affect our health. One way that these technologies could positively impact health is by reducing the use of antibiotics in meat production. According to the World Health Organization, the overuse of antibiotics in farming is a major contributor to the rise of antibiotic resistance, which poses a significant threat to human health. By producing meat in a sterile lab environment, lab-grown meat eliminates the need for antibiotics, which could help to slow the spread of antibiotic resistance. Vertical farming can also have a positive impact on human health by increasing access to fresh produce. In many urban areas, fresh fruits and vegetables are hard to come by, leading to poor nutrition and health outcomes. By growing produce in vertically stacked layers in urban areas, vertical farms can increase access to fresh and nutritious food. These farms can also reduce the environmental impact of traditional farming methods, such as the use of fertilizers and pesticides. Another way that food production impacts human health is through the use of additives and preservatives. Many processed foods contain additives that can have negative health effects, such as artificial colors, flavors, and preservatives. By producing lab-grown meat and growing produce in a controlled environment, it may be possible to eliminate the need for many of these additives. The use of high-pressure processing and other innovative food preservation techniques can extend the shelf life of food without the need for traditional preservatives.

It is important to consider the potential negative impacts of these technologies on human health as well. For example, lab-grown meat may lack certain nutrients that are present in traditionally raised meat, and it is unclear how the body will react to consuming lab-grown meat over the long term. Vertical farming relies heavily on hydroponic systems, which may require more energy and resources than traditional farming methods. If these systems are not powered by renewable energy sources, they could contribute to greenhouse gas emissions and other environmental problems. There are social and cultural impacts to consider. For many people, the act of farming and raising animals is deeply ingrained in cultural and familial traditions. While the use of lab-grown meat and vertical farming may offer many benefits, it is important to consider how these changes will affect cultural practices and traditions related to food production. Innovations in food production have the potential to greatly impact human health, both positively and negatively. It is important to carefully consider the potential benefits and drawbacks of these technologies and to work towards implementing them in a way that maximizes health while minimizing negative impacts. As we move towards a more sustainable and equitable food system, we must keep human health at the forefront of our considerations and work towards creating a food system that is beneficial for both people and the planet. Both lab-grown meat and vertical farming are two innovative methods to enhance food production. Lab-grown meat, also known as cultured meat, is a type of meat that is grown in a lab environment, without the need for animal slaughter. Vertical farming is a technique used to grow crops vertically in a controlled indoor environment. Both innovations have drawn significant attention from the scientific community and the society at large, and the supports for these promising technologies are increasing. Nowadays, the demand for meat products, particularly beef, pork, and chicken, is continuously increasing in conjunction with the global population growth rate. Coupled with the limited resources of land, water, and energy, the traditional livestock farming has become unsustainable. Factory farming has severe ethical and environmental concerns. The development of lab-grown meat has been welcomed as a potential solution to these challenges. The process of growing meat in a lab is both affordable and sustainable due to lower resource requirements. Besides, lab-grown meat has the potential to reduce the greenhouse gas emitted from traditional meat production by up to 96%, lower the use of water by up to 96%, and reduce land use by an estimated 99%.

Lab-grown meat has enormous potential for animal welfare, as it eliminates the need for animal slaughter. The process to produce lab-grown meat requires a small sample of animal cells, which can be collected harmlessly using a biopsy. These cells are placed in a lab-made environment, where they are coaxed to reproduce and grow into the final product. There are no animals involved in the process, and the final result looks and tastes similar to regular meat. Although the production cost is currently high, it is expected to decline significantly as this technology becomes mainstream. The rise of vertical farming, on the other hand, is changing the way we perceive traditional crop production. This innovative farming approach uses stacking layers of plants in a controlled indoor environment that favor high crop yields and low resource usage. The vertical farming method has many benefits compared to traditional farming. For instance, vertical farming requires less water, not only due to the closed-loop irrigation system but also because of the ideal growing conditions in which the plants thrive. This technology also uses less land while yielding more crops compared to traditional farming. Fewer pesticides and herbicides are needed in a vertical farm due to the controlled environment. Another huge feature of vertical farming is the benefits it can bring to urban farming and making agriculture more accessible to urban dwellers. Urban farming can aid in mitigating the issue of food deserts, where some areas are not served by grocery stores that sell fresh fruits or vegetables. Besides providing fresh produce to urban communities, urban farming can also increase social interactions and enhance community engagement. The controlled environment of vertical farms also makes it much easier to transact with retailers, restaurants, and other food-related businesses to expand the market of the farm products. The vertical farming technology facilitates the growth of a wide range of crop varieties, some of which are not traditionally grown in certain regions due to unfavorable climatic conditions. The technology makes it possible to cultivate products like those that are typically imported and typically have a high carbon footprint, such as tomatoes and strawberries, close to where they will be sold. One hindrance to the successful implementation of vertical farming is the need for reliable and affordable energy sources. Vertical farms require artificial lighting and sophisticated irrigation, which often requires substantial amounts of energy. They also require a stable temperature and humidity levels to offer a controlled growing environment. Some new innovations, such as the integration of greenhouses and renewable energy systems, offer feasible solutions in mitigating these challenges; nevertheless, vertical farming must be made more energy-efficient to become more affordable and optimistic. The innovations in food production, specifically lab-grown meat and vertical farming, are projected to have far-reaching impacts on how we produce food in the future. These innovations have enormous potential to increase food security in a time when traditional farming practices are becoming less tenable. While lab-grown meat technology is still relatively new and expensive, it has demonstrated an ability to mitigate the ethical and environmental impacts of traditional meat production. Horizontal farming, on the other hand, can mass-produce crop yields in a more space-efficient manner, with reduced water and land usage and improved growing conditions. Although both innovations require significant investments, their eventual commercialization will undoubtedly create new avenues for feeding an increasingly complex and growing planet.

III. LAB-GROWN MEAT

One key innovation in food production that has been gaining momentum in recent years is lab-grown meat. Also known as cultivated meat, cell-based meat, or clean meat, lab-grown meat is produced by culturing animal cells in a lab rather than raising and slaughtering animals for meat. The process involves taking stem cells from an animal, such as a cow, chicken, or pig, and growing them in a nutrient-rich solution to form muscle tissue. This tissue is then harvested, processed, and turned into meat products, such as burgers, sausages, and steaks, that are virtually indistinguishable from conventionally produced meat.

The idea of producing meat without the need for animal slaughter has been around for several decades but has only recently become a viable option due to advances in technology, increased demand for sustainable food systems, and concerns about the environmental and ethical implications of conventional meat production. One of the main benefits of lab-grown meat is that it requires far less land, water, and feed than traditional livestock farming, and produces significantly lower greenhouse gas emissions. According to a study by the Good Food Institute, producing a pound of beef from cultured cells could require up to 96% less land, 45% less energy, 99% less water, and up to 96% less greenhouse gas emissions than conventional beef production. This makes it a potentially more sustainable and environmentally friendly alternative to conventional meat production.

Another benefit of lab-grown meat is that it could potentially improve animal welfare. Traditional methods of meat production often involve intensive confinement, mutilation, and overcrowding of animals, which can cause physical and psychological stress and lead to a range of health problems. By eliminating the need for animal slaughter and farming, lab-grown meat could reduce the demand for these inhumane practices and provide consumers with a more ethical and humane option for meat consumption. Despite the advantages of lab-grown meat, there are also several challenges that need to be overcome before it can become a mainstream food option. One of the main obstacles is the cost of production, which is currently prohibitively high compared to conventional meat. According to some estimates, lab-grown meat could cost anywhere from $50 to $100 per pound, although some companies are working to bring costs down through innovations in technology and scaling up production. Another challenge is regulatory approval, as lab-grown meat is a relatively new and untested food product that will require approval from government agencies such as the FDA and USDA before it can be sold in stores. There may be public skepticism and resistance to the idea of consuming meat that is produced in a laboratory rather than from a live animal.

Despite these challenges, the market for lab-grown meat is projected to grow significantly in the coming years as more companies enter the field and refine their production methods. Some of the leading companies in the lab-grown meat space include Memphis Meats, Mosa Meat, and Aleph Farms, each of which are working to develop new and innovative ways to produce meat products using cell-based technology. Some of the products that have already been produced and tested include beef burgers, chicken nuggets, and pork sausages, which have been well received by taste testers and food critics alike for their texture, flavor, and nutritional profile.

In addition to the environmental and ethical benefits of lab-grown meat, there are also potential health benefits associated with consuming meat produced using this method. For example, lab-grown meat could be produced without the use of antibiotics or hormones, which are commonly used in conventional meat production and can pose a range of health risks to consumers. Lab-grown meat could be tailored to meet specific nutritional needs or preferences, such as reducing saturated fat content or increasing levels of certain nutrients. Lab-grown meat represents a major innovation in the food industry that has the potential to revolutionize the way we produce, consume, and think about meat. While there are still challenges to overcome before it can become a mainstream food option, the benefits of lab-grown meat are significant and include reduced environmental impact, improved animal welfare, and potential health benefits. As the technology improves and costs come down, it is likely that we will see more and more lab-grown meat products on the market, leading to a shift in the way we eat and produce meat in the future.

DEFINITION AND SCIENTIFIC PROCESS BEHIND LAB-GROWN MEAT

Lab-grown meat, also known as cultured meat or in vitro meat, is a type of meat produced through cell culture. The scientific process behind lab-grown meat involves taking a small sample of animal cells, typically muscle cells, and placing them in a nutrient-rich environment in a lab dish. The cells then begin to multiply and grow, eventually forming muscle tissue that can be harvested and used as meat. The process is similar to how human tissues are grown for medical purposes, such as skin grafts. The concept of lab-grown meat has the potential to revolutionize the way we eat by providing a sustainable and ethical alternative to traditional meat production. It also has the potential to address the environmental and ethical concerns associated with conventional farming and livestock raising. One major advantage of lab-grown meat is its potential to reduce greenhouse gas emissions and other environmental impacts associated with traditional meat production. According to a report by the Intergovernmental Panel on Climate Change, livestock farming accounts for approximately 14.5% of global greenhouse gas emissions. These emissions come from a range of sources, including the methane produced by cows during digestion, the energy required to produce and transport animal feed, and the fossil fuels used in the production and transportation of meat. By contrast, lab-grown meat requires significantly less land, water, and energy to produce, and it produces fewer greenhouse gas emissions. A study from the University of Oxford found that lab-grown meat could potentially reduce greenhouse gas emissions by up to 96% compared to traditional farming methods.

Another advantage of lab-grown meat is its potential to reduce animal suffering and improve animal welfare. Traditional meat production involves raising animals in often cramped and un-sanitary conditions, where they may be subject to stress, dis-ease, and injury. This is particularly true for factory farming, which is the dominant method of meat production in many coun-tries. By contrast, lab-grown meat involves no animal slaughter and requires only a small sample of animal cells to produce. This could potentially reduce the number of animals raised and killed for meat, and it could provide a more humane alternative to conventional meat production. There are also some challenges associated with lab-grown meat that need to be addressed be-fore it can become a viable alternative to traditional meat pro-duction. One major challenge is the cost of production. Currently, lab-grown meat is significantly more expensive to produce than traditional meat. This is due in part to the high cost of the nu-trient-rich growth medium used to culture the cells, as well as the cost of the specialized equipment and facilities required for cell culture. Scientists and entrepreneurs are working to bring down the cost of lab-grown meat through technological innova-tions and economies of scale. In the near future, it is likely that the cost of lab-grown meat will decline as technology advances and the industry matures. Another challenge associated with lab-grown meat is public acceptance and regulation. Many peo-ple are skeptical about the idea of lab-grown meat, and some worry about the safety and health implications of consuming meat that was grown in a lab. There are currently no regulatory frameworks in place for lab-grown meat, which means that the industry operates in a legal gray area. As the industry grows and becomes more mainstream, it will be important to develop clear regulations and standards for the production, labeling, and sale of lab-grown meat to ensure that it is safe, healthy, and transparent for consumers. Lab-grown meat has the potential to revolutionize the way we eat by providing a sustainable and ethical alternative to traditional meat production. The scientific process behind lab-grown meat involves taking a small sample of animal cells and culturing them in a nutrient-rich environment in a lab dish. Lab-grown meat has several advantages over traditional meat production, including its potential to reduce greenhouse gas emissions, animal suffering, and land use. There are also some challenges associated with lab-grown meat, including its high cost of production and the need for regulation and consumer acceptance. Despite these challenges, it is likely that lab-grown meat will become an increasingly important part of the global food system in the coming decades, providing a viable and ethical alternative to conventional meat production.

THE BENEFITS OF LAB-GROWN MEAT COMPARED TO TRADITIONAL MEAT

One of the most significant benefits of lab-grown meat compared to traditional meat is its potential to reduce the environmental impact of meat production. Traditional meat production is incredibly resource-intensive, requiring vast amounts of water, land, and feed. In contrast, lab-grown meat production is significantly more efficient with considerably less environmental impact. One study has suggested that lab-grown meat could require 99% less land, 96% less water, and generate 96% fewer greenhouse gas emissions than traditional meat production.

Lab-grown meat could help reduce the amount of deforestation that occurs to make way for livestock grazing and feed production. Deforestation is a significant contributor to climate change, as it releases greenhouse gases and reduces the amount of carbon dioxide that can be absorbed by forests. The Food and Agricultural Organisation of the United Nations states that the livestock sector is responsible for 14.5% of global anthropogenic greenhouse gas emissions and is a significant driver of deforestation. Lab-grown meat could be one solution to reducing this impact, as it does not require the vast tracts of land used for traditional livestock farming. Another benefit of lab-grown meat is that it could improve animal welfare. Traditional meat production is often associated with poor animal welfare conditions, such as cramped living spaces and inhumane slaughter practices. In contrast, lab-grown meat is produced using animal cell culture technology, meaning that no animals are kept in poor welfare conditions. Lab-grown meat eliminates the need for animal slaughter, significantly reducing animal suffering.

Lab-grown meat also has the potential to reduce the risk of foodborne illness. Traditional meat production is associated with a higher risk of contamination from harmful bacteria such as E. coli and Salmonella. A study by the Centers for Disease Control and Prevention found that foodborne illness is responsible for an estimated 48 million illnesses, 128,000 hospitalizations, and 3,000 deaths in the United States each year. In contrast, lab-grown meat is produced in a controlled environment, reducing the risk of bacterial contamination. As global meat consumption continues to rise, traditional meat production is becoming increasingly unsustainable. Traditional meat production requires vast amounts of feed, much of which is produced using intensive farming practices that contribute to soil degradation and chemical pollution. Traditional meat production is a significant contributor to the depletion of global fish stocks as fish is often used to feed livestock. Lab-grown meat, on the other hand, could be produced using alternative and more sustainable feed sources, such as algae or seaweed. In addition to its benefits, lab-grown meat also presents some challenges that must be addressed to be commercially viable. One of the most significant barriers to the widespread adoption of lab-grown meat is its high production cost. Although the cost of producing lab-grown meat has decreased significantly over the years, it remains an expensive process. As a result, lab-grown meat is not yet commercially viable for mass consumption. As more investment is poured into the research and development of lab-grown meat, it is expected that the cost of producing it will decrease, making it more accessible to the general public. Another challenge for lab-grown meat is public perception. There is still some reluctance among consumers to accept lab-grown meat as a viable and safe alternative to traditional meat. Some people view it as unnatural or as a kind of "frankenfood." Surveys have shown that attitudes towards lab-grown meat are changing, with younger generations being more open to the idea of consuming lab-grown meat. As education and awareness about lab-grown meat continue to increase, it is anticipated that public perception will become more positive. Lab-grown meat has the potential to revolutionize the way we produce and consume meat, addressing many of the environmental, ethical, and health concerns associated with traditional meat production. While there are still some challenges to overcome, such as high production costs and public perception, technological advancements and increased investment are likely to drive the commercial viability of lab-grown meat in the future. Lab-grown meat and other innovations in food production, such as vertical farming, offer exciting possibilities for a more sustainable and equitable food system in the years to come.

THE POTENTIAL DRAWBACKS OF LAB-GROWN MEAT

While there are numerous potential benefits to lab-grown meat, there are also some drawbacks to consider. One significant concern is the possible adverse impacts on the environment. Even though lab-grown meat could reduce the carbon footprint and minimize the need for land and water resources in comparison to traditional animal farming, it still requires significant amounts of energy and resources to produce. The production of lab-grown meat requires a tightly controlled and expensive environment, contributing to the high production costs. The production of meat cells requires the use of chemical substances and antibiotics which could potentially lead to environmental hazards and contributes to the issue of antibiotic resistance. Despite the potential to address some food-related challenges, lab-grown meat could pose other tests and even create more significant problems. For example, it could create economic dilemmas for the meat industry since the cost required for producing lab-grown meat is very high and could result in job losses. Another drawback of lab-grown meat is related to the consumer perception of the product. It is possible that customers recoil from the idea of eating meat that has been developed in a scientific laboratory instead of on a farm. Even some vegans and vegetarians may be hesitant to try the lab-grown meat since it is still a product derived from animal cells and may still be perceived as a product of animal exploitation and cruelty. Consumers may not be able to assimilate the artificiality of lab-grown meat, and they may perceive it as being 'unnatural' or unappetizing. The nutritional value of lab-grown meat is not yet fully understood, and public concerns about the health impact of consuming products derived from laboratories have been raised. Some studies suggest that lab-grown meat products carry some significant nutritional advantages but may have unintended consequences like allergies or other health risks. As a result, the approval of the safety and quality of lab-grown meat by governments and official international organizations is critical before it can be successfully marketed. Another drawback is the ethics involved in lab-grown meat production, which could lead to the loss of animal welfare value. The premise of lab-grown meat as a more humane way of rearing animals may not hold since it is derived from the same cell of an animal, and researchers suggest that some animals used for tissue culture may have been subjected to animal cruelty and violation. It is uncertain if the method of lab-grown meat production actually increases animal welfare since, in its inception, it relies heavily on the use of animals as cell donors. Also, there are concerns from activists about the misuse of genetic engineering technologies to create frankenmeat, which could precipitate unforeseen health risks if not adequately regulated and tested. The lack of public awareness and acceptance represents another major drawback to lab-grown meat, especially in underdeveloped countries. Lab-grown meat requires expensive equipment and is demanded by a mostly advanced middle-class segment of society. This could exacerbate the already existing inequalities in a society, where only a few people have access to the healthier, more sustainable, and safer food option that lab-grown meat promises. This could generate enormous tension between the "haves" and "have-nots" in the wider society. A side-effect of lab-grown meat production is the potential for failures in the supply chain, which could result in the entire company's disaster. Even though the industry is still in its infancy, laboratory-grown meat has already attracted some industry actors, both local and foreign, ready and willing to invest in the new venture. Without a proper supply chain management structure, it is likely that difficulties in production caused by technical or logistics challenges could impede the success of the industry. The last significant drawback of the lab-grown meat can merely stem from a simple portion of customer preference. Many customers may continue to prefer the taste and texture of traditional meat products that come from animals. Taste is an essential factor that drives the customers' demand and loyalty, and it is still unknown how well lab-grown meat will be received by the customers. It is essential to have a robust and reliable system to guarantee flavor and consistency to a new system without animal genetics. Lab-grown meat represents a potential revolution in food production, offering hope in addressing many of the environmental, health, and ethical issues facing the meat industry. This technology is still in its beginning stages, and while promising, it would be wise to seriously weigh the benefits and drawbacks before embracing it fully. The drawbacks highlighted earlier serve as a reminder that there is still a lot of work to be done, and careful evaluation and regulation are necessary to ensure that this emerging industry is both sustainable and that it meets the demands of the modern consumer. In the end, it may be ultimately up to the customer and the market to decide whether lab-grown meat makes it to the mainstream as consumers decide on the health, ethics, nutritional value, and environmental impact of their food choices. There is no denying that the way we produce and consume food is changing. We have seen a surge in interest in the development of lab-grown meat and vertical farming. These innovations offer potential solutions to the growing concerns surrounding the environmental impact of traditional food production methods, as well as ethical concerns around animal welfare. The development of lab-grown meat, also known as cultured meat, involves the production of meat in a laboratory from animal cells, rather than breeding and raising animals for food. This technology has the potential to eliminate the need for animal slaughter altogether, which would alleviate ethical concerns while also offering a more sustainable method of food production. Similarly, vertical farming can address concerns about land use and environmental degradation associated with traditional agriculture. This technique allows for the growth of crops in a controlled environment, such as a skyscraper or warehouse, using technology such as hydroponics and aeroponics. Lab-grown meat has been in development for several years now, and although it is still not commercially available, it has come a long way. The potential benefits of lab-grown meat are many. Since it is produced in a lab, it is free from antibiotics and other substances that are commonly used in traditional animal agriculture, making it a healthier option for consumers. The production of lab-grown meat would not require the extensive use of land and water resources that is currently necessary for traditional meat production. This would significantly reduce the environmental impact of meat production, which is currently a major contributor to greenhouse gas emissions.

There are still some challenges facing the implementation of lab-grown meat on a large scale. One major hurdle is cost. Currently, the process of producing lab-grown meat is very expensive, and it would need to become more affordable before it can be widely adopted. There is still significant work to be done in terms of taste and texture. While some consumers have reported that lab-grown meat tastes like traditional meat, others have noted that it has a different texture. Until these issues are resolved, there may be resistance to adopting lab-grown meat as a viable alternative to traditional meat. Vertical farming, on the other hand, has already become a reality in some parts of the world. This technique involves growing crops in a controlled environment, such as a skyscraper or warehouse, using hydroponics or aeroponics—methods that allow plants to grow without soil. This type of farming has many potential benefits over traditional agricultural methods. For one, it requires significantly less land, water, and pesticides, making it a more environmentally sustainable option than traditional farming. Since the crops are grown indoors, they are not at risk of being damaged by extreme weather conditions or pests. This makes vertical farming a more reliable method of food production, which is especially important in areas that are prone to droughts or other environmental challenges. Another potential benefit of vertical farming is that it allows for year-round crop production. Since crops are grown in a controlled environment, they can be grown year-round, eliminating the need to wait for the right season to plant and harvest in traditional agriculture. This can significantly increase the amount of food produced per unit of land, which is critical given the growing global population and the increasing demands for food. There are also some challenges facing the widespread implementation of vertical farming. One of the most significant is the high cost of setting up and maintaining a vertical farm. The technology required for this type of farming is still relatively new and expensive, which makes it difficult for farmers to adopt on a large scale. There are concerns around the energy use associated with vertical farming. Since the crops are grown indoors, they require additional lighting and heating, which can be energy-intensive. The innovations in food production, such as lab-grown meat and vertical farming, have the potential to revolutionize the way we eat. They offer promising solutions to the challenges facing traditional methods of food production, including environmental degradation and animal welfare concerns. There are still significant challenges facing the widespread adoption of these technologies. Addressing these challenges will require continued research, development, and investment. Only then can we hope to realize the full potential of these innovations and create a more sustainable and equitable food system for all.

IV. THE ENVIRONMENTAL IMPACT OF LAB-GROWN MEAT

While lab-grown meat might offer an alternative to traditional animal agriculture, it is not without its environmental costs. One important consideration is the energy input required to produce cultured meat. Developing a viable and sustainable meat alternative without a significant impact on the planet and resources is the ultimate goal of scientists, and they are still in the process of perfecting the technology. Currently, the production of cultured meat requires a lot of energy for heating, sterilizing, and maintaining a sterile environment, which often comes from non-renewable sources. Although the energy consumption is lower than traditional meat production, it still has considerable environmental costs. Researchers are already working on reducing the energy consumption and exploring alternative energy sources to make lab-grown meat a more sustainable option for meat production. Another important environmental consideration is the use of resources such as land, water, and feedstock. While lab-grown meat production does not require the use of land, it still needs a considerable amount of water and feedstock to sustain the cell cultures. The production of lab-grown meat requires nutrients for the cells, and they are often supplied by agricultural crops. Researchers need to find sustainable and efficient ways to grow the crops needed to provide the nutrients for cell culture. Since water is becoming a scarce resource, it is crucial to ensure that the production process of lab-grown meat is water-efficient and does not compete with other important uses of water. It is necessary to consider the total environmental impact of the production of lab-grown meat and ensure that it is a sustainable and responsible alternative. The environmental impacts of lab-grown meat also extend to the issue of waste. Lab-grown meat production generates a considerable amount of waste, from the unused nutrients for cell-culture to the lefto-ver materials from the production process. Disposing of this waste is not only expensive, but it also poses environmental haz-ards. Researchers are working on developing bio-based solutions to tackling this problem. For example, they are investigating the possibility of using biodegradable materials that can be broken down without releasing harmful toxins or pollutants into the en-vironment. There is the issue of the carbon footprint. While the carbon footprint of lab-grown meat is still lower than that of traditional meat, there is still room for improvement. Research-ers are exploring ways to reduce the carbon footprint further by adopting sustainable and renewable energy sources, using effi-cient production techniques, and developing bio-based solutions to minimize waste and pollution. A study showed that the com-plete replacement of traditional meat products with lab-grown meat would lead to a 96% reduction in greenhouse gas emissions. The adoption of lab-grown meat can potentially play a signifi-cant role in mitigating climate change. While lab-grown meat offers an exciting alternative to traditional meat production, it is still a new technology that needs to evolve over time to be-come a practical and sustainable solution. Researchers need to consider the environmental impacts of lab-grown meat produc-tion and weigh the benefits and trade-offs against traditional meat production. It is essential to ensure that any solution to the problem of meat production is sustainable, ethical, and does not leave any significant impact on the environment. Both lab-grown meat and vertical farming offer innovative solutions to the problem of food production and will likely change the way we eat in the future. Lab-grown meat has the potential to reduce the number of animals slaughtered for meat, while vertical farming can provide fresh and locally grown produce year-round. Both technologies have the potential to address the challenges of feeding the world's growing population in a sustainable and responsible manner. These innovations still have their challenges, and researchers need to continue working on developing more efficient and sustainable techniques. While both lab-grown meat and vertical farming might not be the ultimate solution to the problem of feeding the world, they are a good step in the right direction towards creating a more sustainable and equitable food system. It is our responsibility to invest in such innovations to realize their full potential and to ensure that everyone has access to healthy, nutritious, and affordable food.

THE ROLE OF LAB-GROWN MEAT IN REDUCING GREENHOUSE GAS EMISSIONS

The United Nations' Food and Agriculture Organization (FAO) estimated that meat production is responsible for 14.5% of total global greenhouse gas emissions. It is because animal agriculture is linked to various environmental problems, including deforestation, water pollution, and greenhouse gas emissions, among others. This issue has been a subject of concern for climate activists in recent years, as we are experiencing the effects of global warming and climate change firsthand. In this regard, lab-grown meat is seen as a potential solution to this problem. One of the main reasons why meat production is problematic for the environment is that livestock requires a lot of resources to produce. These resources range from water, land, and feed to energy. According to research conducted by the Good Food Institute, lab-grown meat production has the potential to significantly reduce the demand for these resources. For instance, cultured meat requires 90% less water and land compared to traditional animal agriculture. It produces 78% less greenhouse gas emissions than conventional meat production. In comparison, the carbon footprint of one kilogram of beef is equivalent to driving a car for 200 kilometers. If scaled up and fully adopted, this emerging technology could have a significant positive impact on the environment. In addition to its low carbon footprint, lab-grown meat technology reduces reliance on traditional farming practices that often cause deforestation and ecosystem degradation. The FAO reports that 70% of global deforestation is driven by the expansion of agricultural land-use, with livestock production as the main driver. When implemented on a large scale, lab-grown meat technology has the potential to disrupt this trend by reducing the need for vast tracts of land and grazing areas. Similarly, reduced demand for feed crops (e.g., corn and soy) would put less pressure on terrestrial ecosystems by allowing for more land to remain in its natural state. Lab-grown meat has the potential to solve the problem of overfishing and unsustainable aquaculture practices since it can be produced without fishing for wild animals or using feed that causes excessive environmental damage. As we navigate the complex and interconnected challenges of the 21st century, lab-grown meat may hold the key to securing our food supplies. Global food security is already a challenge, but it is exacerbated by the inefficiencies inherent in traditional animal-based food production systems. Factors, such as supply chain disruption, limited growing seasons, and foodborne disease outbreaks, create uncertainties in food availability and limit access to diverse, healthy, and cost-effective diets. With the development of lab-grown meat, however, we move closer to developing a future food system that can ensure predictable and sustainable food supplies in the face of unforeseen challenges. It is important to note that lab-grown meat is not entirely without its challenges and critics. While laboratory-grown meat is generally considered safe to consume, questions of its nutritional value and production costs remain. Animal rights and ethical considerations, such as concerns over animal suffering and genetic engineering, have been raised by critics of the technology. The use of genetically modified organisms (GMOs) in this emerging food industry is also a point of concern as the long-term health and environmental effects of GMOs are still being studied. Despite these challenges, the potential benefits of lab-grown meat technology cannot be ignored. The scientific advancements in this emerging industry have led to a growing number of startups, investors, and consumers interested in lab-grown meat. Given the urgency of the climate crisis and the need for a food system that is both sustainable and equitable, it is likely that we will see further development and adoption of this technology in the future. It is worth considering the impact that lab-grown meat technology will have on society as a whole. Apart from the environmental and economic benefits, lab-grown meat technology also offers an opportunity to address social and cultural tensions surrounding meat consumption. For instance, there are growing concerns about the consumption of meat and its link to health challenges such as obesity and various chronic diseases. There are ethical concerns related to the treatment of livestock animals in animal agriculture. By providing a meat alternative that is free of these concerns, lab-grown meat technology offers a promising alternative to consumers who are increasingly looking for more sustainable and ethical food choices. The continued development and adoption of lab-grown meat technology has the potential to provide a more sustainable, secure, and equitable food system. The technology addresses several environmental challenges, including the reduction of greenhouse gas emissions, resource use, and deforestation. It has the potential to address social and cultural challenges related to animal welfare and public health. While concerns and critics remain concerning the technology, the potential benefits of developing and adopting lab-grown meat technology are too important to ignore.

THE SUSTAINABILITY OF LAB-GROWN MEAT IN TERMS OF LAND AND WATER RESOURCES

The sustainability of lab-grown meat in terms of land and water resources is one of the most critical aspects of this technology's future. It is estimated that the world population will reach 8 billion by 2025, and it is projected that food production will need to increase by up to 70% to feed the growing population. The shift towards plant-based diets is also gaining momentum due to environmental and health reasons. In this context, lab-grown meat offers a viable alternative that can meet the growing demand for meat without putting excessive pressure on already-strained land and water resources. To understand the sustainability of lab-grown meat, it is essential to compare it with traditional meat production. Traditional animal agriculture practices require vast amounts of land to grow crops to feed animals, house animals, and dispose of waste. The United Nations' Food and Agricultural Organization (FAO) estimates that about 30% of the world's land surface is used for livestock rearing. The production of meat requires much more water than the production of plant-based foods. According to the FAO, it takes up to 15,500 liters of water to produce one kilogram of beef, while it takes only about 250 liters of water to produce one kilogram of potatoes. In contrast, lab-grown meat production requires significantly less land and water. The process involves growing muscle tissue cells in a lab setting, which can be done using small amounts of land and water. According to Memphis Meats, a leading company in lab-grown meat production, their process requires 90% less land and 90% less water than traditional meat production. Because the process is not reliant on animal agriculture practices, it eliminates the environmental impact of animal waste, which is a leading cause of water pollution.

Another aspect of lab-grown meat's sustainability is its potential to reduce greenhouse gas emissions. Traditional animal agriculture practices are responsible for a significant portion of global greenhouse gas emissions. It is estimated that animal agriculture accounts for about 18% of global greenhouse gas emissions, more than the entire transportation sector. The production of lab-grown meat produces significantly less greenhouse gas emissions than traditional meat production. According to one study, produced by the think-tank RethinkX, lab-grown meat could reduce greenhouse gas emissions by up to 96%, relative to traditional meat production. The sustainability of lab-grown meat depends on the energy source used to power the lab-grown meat production process. The process of growing meat in a lab setting requires a lot of energy, primarily in the form of electricity. If the electricity used to power lab-grown meat production comes from non-renewable sources, it could offset the environmental benefits of this technology. If the energy used to power lab-grown meat production comes from renewable sources such as wind, solar, or geothermal, then lab-grown meat could potentially be a carbon-neutral food source. Lab-grown meat may also help to alleviate food security concerns. Lab-grown meat production can take place anywhere, regardless of climate, geography, or soil quality. It enables food production to be decentralized, providing communities with the ability to grow fresh meat without the need for traditional animal agriculture practices or transportation. This could be especially important for communities that lack access to nutritious meat, such as those in remote or drought-stricken areas. The sustainability of lab-grown meat, in terms of land and water resources, is promising. It requires significantly less land and water than traditional meat production, and it eliminates the environmental impact of animal waste, reducing the water pollution. It can potentially reduce greenhouse gas emissions by up to 96%. Nonetheless, one of the challenges of this technology will be to ensure that the energy used to power the production process comes from renewable sources. If lab-grown meat becomes part of the solution to a more sustainable food production system, it could potentially have a significant impact on the way we eat. It may help to alleviate food insecurity, reduce the pressure on already-strained land and water resources, and mitigate the environmental impact of traditional meat production practices.

THE ETHICAL IMPLICATIONS OF LAB-GROWN MEAT PRODUCTION

While technology continues to advance, our relationship with animals and their place in our food system is a topic of ongoing debate. The ethical implications of lab-grown meat production, or "cultured meat," are complex and far-reaching. At their core, the ethical concerns are about whether it is ethical to create and consume a product that is technically meat but has never had a brain or a physical form beyond muscle tissue. Some argue that lab-grown meat production will decrease animal suffering and environmental degradation, making it an ethical alternative to traditional animal agriculture. Others believe that cultured meat production is simply commodifying animals and that dismantling animal agriculture entirely is the only way forward. One of the primary ethical concerns surrounding lab-grown meat production is related to the use of animal cells to create the meat. Although no animals are harmed in the production of lab-grown meat, the cells required to make it must initially come from an animal. These cells are usually collected through a biopsy or some other non-harmful method. Some groups argue that the use of animal cells, even without harm, constitutes an unethical use of animals as a resource. For example, animal rights groups such as PETA (People for the Ethical Treatment of Animals) argue that the ethical alternative to animal agriculture is simply not using animals at all, rather than simply changing the methods used to exploit them. Thus, from this perspective, creating cultured meat products is just another form of commodification of animals, which is itself unethical.

Another concern relates to the possible environmental and health implications of lab-grown meat production. Some advocates argue that the mass production of cultured meats will significantly reduce environmental harms associated with traditional animal agriculture, such as deforestation, greenhouse gas emissions, and the depletion of water resources. Yet, some argue that we may be trading one set of problems for another by relying on science and technology as a solution. For example, possible negative environmental impacts of mass-producing lab-grown meat include the use of large amounts of energy and materials for tissue culture, disposal of waste products, and the possible increase in industrialized agriculture practices. There is a potential for synthetic meat to distance ourselves further from the realities of animal-based consumption, potentially weakening our moral conscience towards animal welfare. Thus, we must still remain vigilant to ensure that our relationship with animals, and how we produce and consume food, is ethical and humane. The ethical concerns surrounding lab-grown meat production are complex and multi-layered. While it is true that cultured meat could reduce animal suffering, it is also true that the process of creating it raises ethical issues of its own. Within the context of animal welfare, approaches to agriculture ranging from wildly different, yet proposed methods such as deep sustainable to veganism present a conundrum that speaks to fundamental differences in immanent ethics. To get a better idea of how animal welfare and ecological conservation may create the ethical framework for lab-grown meat production, it is beneficial to compare it to the ethical arguments supporting vertical farming's success.

A COMPARISON TO VERTICAL FARMING

Like lab-grown meat production, vertical farming presents a method for food production that promises significant environmental and efficiency benefits. Proponents of vertical farming argue that it could significantly reduce the ecological footprint associated with traditional agriculture by taking advantage of space vertically instead of spreading out horizontally in traditional fields. When food is produced closer to where it is consumed, there is less need for transportation and refrigeration, reducing the fossil fuel use and CO_2 emissions associated with transportation. Because it does not require the use of soil, this type of farming can be located in urban settings and is not subject to the many variables of traditional farming practices that can cause changes in crop yields or quality. As with lab-grown meat production, there are also ethical considerations associated with vertical farming, principally concerning the quick commodification of large-scale agricultural practices and the continued removal of animals in food systems at-large.

Lab-grown meat production presents both opportunities and challenges. It is possible that it will become the answer to many of the problems associated with animal agriculture, such as animal suffering and environmental degradation, but it is equally crucial that we do not allow the benefits to overshadow the ethical concerns that arise from this production method. A critical approach towards animal welfare and ecological stewardship are just crucial pillars of being responsible citizens of Earth. Continued innovation in food production, such as lab-grown meat and vertical farming, merits ongoing analysis and discussion, from a range of disciplines perspective, to determine collectively how best to navigate such morally complex issues for a sustainable future. Lab-grown meat and vertical farming are two innovative technologies that are poised to revolutionize the way we produce and consume food. While these innovations are still in their early stages, they show tremendous potential for addressing some of the most pressing challenges facing our global food system. Lab-grown meat offers a solution to the many ethical and environmental concerns that arise from our current meat production systems, while vertical farming can provide fresh, healthy produce to urban populations in a sustainable way. These innovations will not only change the way we eat but also have the potential to transform the entire food industry, creating new opportunities for sustainable and ethical food production.

Lab-grown meat, also known as cultured meat, involves the production of meat from animal cells in a lab setting, rather than from living animals. By avoiding the need for animal slaughter, lab-grown meat offers a more ethical alternative to traditional meat production. It has the potential to be much more sustainable than traditional meat production, which is known to be a major contributor to greenhouse gas emissions and deforestation. Lab-grown meat could significantly reduce the environmental impact of meat production by requiring fewer resources to produce, including land, water, and energy. While lab-grown meat is still in its early stages, researchers have made significant progress in developing the technology. In 2013, the first lab-grown burger was produced, costing around $300,000. Since then, the technology has improved, and the cost has gone down significantly. In 2020, Singapore became the first country to approve the sale of lab-grown meat, with more countries likely to follow suit. While the technology is still too expensive for mass consumption, it is expected to become more affordable as it becomes more widely adopted. One of the main benefits of lab-grown meat is that it could fundamentally change the way we think about meat production and consumption. With traditional meat production, the cost and environmental impact of raising an animal means that meat is often consumed in small quantities. Concerns about animal welfare have led many to adopt vegetarian or vegan diets. Lab-grown meat could offer a more sustainable and ethical way to consume meat, potentially increasing the amount of meat we consume without the environmental and ethical concerns associated with traditional meat production. Vertical farming, on the other hand, offers a way to produce fresh produce in urban areas where space is limited. Vertical farms use a combination of LED lights, hydroponics, and other high-tech methods to grow crops indoors, often in buildings that would otherwise be unused. This technology has the potential to dramatically reduce the distance that food needs to travel, reducing carbon emissions associated with transportation. It also offers a way to produce fresh produce year-round, regardless of weather conditions. One of the key advantages of vertical farming is that it can be done without the use of pesticides and other chemicals commonly used in traditional agriculture. By growing crops in a controlled environment, vertical farms can avoid many of the problems associated with traditional agriculture, including disease, pests, and weather-related damage. This means that the produce produced by vertical farms is often healthier and safer for consumption than conventionally grown produce. Another benefit of vertical farming is that it can produce much higher yields per square foot than traditional agriculture. This is because crops can be grown vertically, taking advantage of unused space in the building. Crops can be grown much closer together, reducing the amount of space needed for each individual plant. This means that a single vertical farm can produce a large amount of produce, even in a relatively small space. Vertical farming also has the potential to create new economic opportunities in urban areas. By utilizing unused buildings to produce food, vertical farms can bring new life to downtown areas and create jobs in the local community. The technology used in vertical farming is highly scalable, meaning that it can be adopted in a variety of different settings, from small-scale community gardens to large-scale commercial farming operations. While lab-grown meat and vertical farming represent exciting innovations in food production, they are not without challenges. For lab-grown meat, one of the biggest challenges is the cost of production. While the cost has gone down significantly, it is still too expensive for mass consumption. There are concerns around the safety and long-term health impacts of consuming lab-grown meat. While early studies have shown that it is safe for human consumption, more research is needed to fully understand any potential risks. For vertical farming, one of the main challenges is the high upfront cost of building and equipping a vertical farm. There are concerns around the energy usage of vertical farms, as they require large amounts of electricity to power the LED lights and other equipment. There are concerns around the scalability of vertical farming, as it may be difficult to produce certain crops in a controlled environment. Despite these challenges, lab-grown meat and vertical farming represent two promising solutions to some of the most pressing challenges facing our global food system. As these technologies continue to evolve and improve, they have the potential to fundamentally change the way we produce and consume food. By offering more sustainable and ethical solutions to food production, these innovations could help create a healthier, more equitable, and more sustainable food system for all.

V. LAB-GROWN MEAT AND HUMAN HEALTH

Apart from the ethical and environmental concerns, lab-grown meat presents a range of health benefits that could provide potential solutions to some of the most significant health problems. For instance, since conventional meat is often associated with high levels of saturated fats, cholesterol, and other harmful substances, the idea of lab-grown meat that contains healthy fats and high-quality proteins is indeed an attractive proposition. More importantly, lab-grown meat does not pose the risk of contaminating foodborne illnesses such as E. coli, Salmonella, and other bacteria associated with conventional meat production.

According to the Centers for Disease Control and Prevention (CDC), every year, about one in six Americans - approximately 48 million people - suffer from foodborne illnesses, while 3,000 people die from it. One of the main sources of food contamination is meat, as it can harbor harmful bacteria throughout its production cycle, from animal farming to processing and packaging. Since lab-grown meat would eliminate the need for animal farming and the potential for contamination as a result, the risk of foodborne illnesses would be minimal. Lab-grown meat's nutritional content can be precisely controlled. As such, it provides an ideal solution for many people who have specific dietary requirements. For example, for those with allergies to various types of meat or plant-based diets, lab-grown meat offers a sustainable alternative. It can provide a reliable source of protein for people who do not have access to high-quality meat or who cannot afford it. This benefit is of particular importance in regions of the world where protein sources are limited, or food insecurity is prevalent. Lab-grown meat could have positive implications for public health, given the growing health concerns associated with conventional meat consumption such as antibiotic resistance and zoonotic diseases. According to the World Health Organization (WHO), the overuse of antibiotics in animal farming is contributing to the emergence of antibiotic-resistant bacteria, which pose a significant health threat to humans. Zoonotic diseases, such as Mad Cow disease, avian influenza, and swine flu, have also been associated with conventional meat production, leading to outbreaks that can lead to human deaths. In contrast, lab-grown meat could ensure a safer food supply, free of harmful pathogens and antibiotics. It could also reduce the need for the use of antibiotics in animal agriculture. Accordingly, this could contribute to the realization of containment and prevention efforts, minimize the risk of transmission of zoonotic diseases and antibiotic resistance, and promoting public health. Although lab-grown meat offers several health benefits, it is still a relatively new development that requires further research and testing. One critical challenge that the industry faces is to ensure that all the nutrients required for the growth of meat are supplied in a cost-effective manner. Currently, laboratories are using expensive and energy-intensive processes that make the production of lab-grown meat too costly. Research is ongoing to develop more sustainable processes with more affordable and efficient nutrient media. The industry needs to design measures for the regulation of lab-grown meat production, safety, and labeling to ensure public confidence in and acceptance of the product. The industry must comply with the Food and Drug

Administration's (FDA) safety requirements, which govern the use of animal cells for human consumption. The FDA has not yet approved lab-grown meat for commercial consumption, but regulatory bodies are working on frameworks to ensure that the industry operates safely and meets the required standards.

Lab-grown meat has the potential to revolutionize the food industry. It can provide sustainable, affordable, and healthy alternatives to conventional meat. It could also reduce animal agriculture's environmental impact, as well as offer significant improvements in public health. The industry needs to overcome the significant technical and cost challenges associated with the production of lab-grown meat, and work with regulatory bodies to ensure public safety. Although there has been significant progress made in the development of lab-grown meat, there is still a long way to go before it becomes a viable and practical mainstream alternative to conventional meat. Nevertheless, given the current challenges associated with conventional meat production, the promising outlook for lab-grown meat highlights its potential to shape the future of food production.

THE NUTRITIONAL VALUE OF LAB-GROWN MEAT

One of the most significant benefits of lab-grown meat is its potential to offer more nutritious options for consumers. Unlike traditional animal meat, which can contain high levels of saturated fat, cholesterol, and other potentially harmful compounds, lab-grown meat can be engineered to include healthier amounts of protein and other essential nutrients. In fact, studies have suggested that lab-grown meat could contain higher levels of protein and lower levels of fat than traditional animal meat, making it a healthier alternative for people looking to improve their diets. Since lab-grown meat is produced in a controlled environment, companies are able to regulate the nutrient content of their products, making it possible for them to create meat with specific nutritional profiles tailored to individual consumer needs. One of the primary ways in which lab-grown meat could improve the nutritional value of our food system is by reducing the amount of saturated fat in our diets. Saturated fat is a type of fat that is found in high concentrations in red meat and other animal products and has been linked to an increased risk of heart disease, stroke, and other health problems. Lab-grown meat, however, can be produced with significantly lower levels of saturated fat than traditional animal meat, making it a healthier option for consumers looking to reduce their intake of this harmful nutrient. Lab-grown meat can be engineered to contain more beneficial nutrients, such as omega-3 fatty acids, which are typically found in oily fish like salmon and trout. In fact, some companies are already exploring ways to create lab-grown fish products that could provide these essential amino acids without the risk of mercury contamination that can be associated with wild fish. Another way in which lab-grown meat can improve the nutritional value of our food system is by addressing the issue of micronutrient deficiencies. Micronutrients are essential vitamins and minerals that our bodies require in small amounts to maintain optimal health, but which can be difficult to obtain in sufficient quantities through a typical Western diet. For example, iron deficiency is a common problem in many parts of the world, particularly among women and children, and can lead to a range of health problems such as anemia and fatigue. By carefully controlling the nutrient content of their products, companies producing lab-grown meat and other foods can help to address these deficiencies and provide essential vitamins and minerals that are currently lacking in many people's diets.

Lab-grown meat could also offer a more sustainable option for consumers looking to improve the nutritional value of their diets. Traditional animal agriculture is a major contributor to environmental problems such as deforestation, water pollution, and greenhouse gas emissions, and is unsustainable in the long term. By contrast, lab-grown meat can be produced using significantly less land, water, and other resources, and has the potential to reduce the environmental impact of our food system. This could have important benefits for both human health and the health of the planet, making it a win-win solution for consumers, producers, and the environment. Of course, it's worth noting that lab-grown meat is not a panacea for all of the problems facing our food system. There are still many questions about the safety, efficacy, and long-term impact of this technology, and it will likely be some time before we see widespread adoption of lab-grown meat products in supermarkets and restaurants. There are still concerns about the potential risks of genetic engineering and the safety of lab-grown meat for human consumption.

Despite these challenges, however, the potential benefits of lab-grown meat and other innovations in food production are significant. By tapping into cutting-edge technologies and scientific research, we can begin to create a more sustainable, healthier, and more nutritious food system for ourselves and for future generations. Whether through lab-grown meat, vertical farming, or other innovations, the future of food promises to be an exciting and transformative era in human history. As we continue to explore these new frontiers of food production, we can look forward to a world where healthy, sustainable, and delicious food is available to all.

THE POTENTIAL RISKS ASSOCIATED WITH LAB-GROWN MEAT CONSUMPTION

While lab-grown meat may offer many benefits to consumers, there are also potential risks associated with its consumption. One of the main risks is that the long-term effects of consuming lab-grown meat are not yet known. As a relatively new technology, there has not been enough time to properly assess the health implications of lab-grown meat, and there may be unforeseen consequences that arise over time. The production of lab-grown meat may require the use of antibiotics and growth hormones, which could lead to the development of antibiotic-resistant bacteria and other negative health outcomes. This could have serious implications for public health, as the overuse of antibiotics in animal agriculture has already led to the proliferation of antibiotic-resistant bacteria. Another potential risk associated with lab-grown meat consumption is that it could have negative environmental impacts. While lab-grown meat has the potential to reduce the environmental impact of traditional animal agriculture, it is not completely clear what the full impact of the technology will be. For example, it may require large amounts of energy to produce lab-grown meat, which would increase greenhouse gas emissions. The use of artificial lighting and other lab equipment may require the use of non-renewable resources. This could lead to unintended consequences, such as increased reliance on fossil fuels and other non-renewable resources, which could exacerbate climate change. There is the issue of the social and ethical implications of lab-grown meat production. Many people may object to the idea of consuming meat that has been grown in a laboratory and may have concerns about the way in which the technology is developed and produced. For example, some may argue that the cells used to produce lab-grown meat should be sourced from natural sources, rather than from animals that have been raised in inhumane conditions. This raises important questions about the ethics of animal agriculture and the role of technology in addressing social and environmental problems. Lab-grown meat and vertical farming are exciting innovations in the field of food production that have the potential to revolutionize the way we eat. They are not without risks. The potential health, environmental, and social implications of these technologies must be carefully considered and monitored to ensure that they are used in a way that promotes sustainable and equitable food systems. The success of these innovations will depend on their ability to address some of the most pressing challenges facing the global food system, from environmental degradation and food insecurity to public health and animal welfare.

THE ROLE OF LAB-GROWN MEAT IN COMBATING FOOD-BORNE ILLNESS

One of the most significant advantages of lab-grown meat is that it can significantly reduce the risk of food-borne illnesses. This is because lab-grown meat is produced under carefully controlled conditions, with no exposure to the pathogens that are commonly found in traditional animal farming environments. In contrast, traditional animal farming is associated with a high risk of food-borne illnesses, such as salmonella and E. coli infections. These illnesses can be extremely dangerous and even fatal, particularly in vulnerable populations such as the elderly, young children, and those with compromised immune systems. Food-borne illnesses can lead to significant economic costs, including lost productivity and medical expenses. Lab-grown meat has the potential to reduce the risk of such illnesses by eliminating the need to raise animals in traditional farming environments. Instead, lab-grown meat is produced in sterile, contained environments where the risk of contamination is minimal. As a result, lab-grown meat can be produced without the use of antibiotics or other potentially harmful chemicals, further reducing the risk of contamination. This is particularly important as the overuse of antibiotics in traditional farming has led to the emergence and spread of antibiotic-resistant bacteria, making it more difficult to treat infections in humans. The safety benefits of lab-grown meat are already being recognized by the food industry, and several companies are actively working to bring it to market. For example, Memphis Meats, a California-based startup, has developed a lab-grown meatball and chicken strip that they plan to release to consumers in the near future. Similarly, Mosa Meat, a Dutch company, has produced a lab-grown hamburger that was taste-tested in 2013. These companies are aiming to meet the growing demand for environmentally friendly and sustainable food options that are also safe and healthy to eat. In addition to its potential to reduce the risk of food-borne illnesses, lab-grown meat may also have other health benefits. For example, it may be lower in saturated fat and cholesterol than traditional meat products, making it a healthier option for consumers. This is significant as a diet high in saturated fat and cholesterol is a known risk factor for heart disease, which is the leading cause of death worldwide. It is important to note that the long-term health effects of lab-grown meat are still unknown, as it is a relatively new technology. As a result, further research is needed on the potential health benefits and risks associated with lab-grown meat before it can be widely adopted. There are ethical concerns associated with lab-grown meat that should be considered. Some have raised concerns about the use of animal cells to produce lab-grown meat, arguing that it still involves the exploitation of animals. Others have raised questions about the long-term sustainability of lab-grown meat, as it requires a significant amount of energy and resources to produce. These concerns should be carefully weighed against the potential benefits of lab-grown meat, and further research should be conducted to address them. Despite these concerns, the potential of lab-grown meat to revolutionize the food industry and improve food safety is undeniable. As the world's population continues to grow, and the demand for meat increases, lab-grown meat offers a sustainable, environmentally friendly, and safe alternative to traditional meat products. It is important that we continue to carefully monitor its development and implementation to ensure that it is safe, ethical, and sustainable.

The future of food is likely to be shaped by a combination of technological innovations, such as lab-grown meat and vertical farming, and changing consumer preferences and behaviors. These technologies offer a promising solution to many of the challenges facing the food industry, from the need to reduce greenhouse gas emissions and improve food safety to the growing demand for sustainable and healthy food options. As these innovations continue to evolve and become more widely adopted, they have the potential to transform the way we eat, creating a more sustainable and equitable food system for all.

One of the most significant innovations in food production that is poised to transform the way we eat is lab-grown meat. Lab-grown meat, also known as cultured meat or clean meat, is produced by using stem cells from animals to grow muscle tissue in a lab. This tissue is then harvested, processed, and packaged for consumption. Lab-grown meat has the potential to revolutionize the food industry in several ways. First, it could help to address the environmental impact of conventional animal agriculture. Livestock farming is a major contributor to greenhouse gas emissions, land degradation, and water pollution. By producing meat in a lab, it would be possible to reduce the need for land, water, and other resources required for conventional animal agriculture. Lab-grown meat could help to address issues related to animal welfare. The meat industry is notorious for its ethical problems, including cramped living conditions and inhumane slaughter methods. By producing meat in a lab, we could avoid these issues entirely and ensure that animals do not suffer needlessly to feed us. Another innovation that could transform the way we eat is vertical farming. Vertical farming is a method of growing crops in vertically stacked layers, using artificial lighting, controlled temperature, and humidity. This approach to farming has several advantages over traditional agriculture. One of the most significant benefits of vertical farming is that it allows for year-round crop cultivation, regardless of seasonal changes or weather conditions. This means that fresh produce can be grown locally and delivered to consumers quickly, reducing food waste and carbon emissions associated with long-distance transportation. Vertical farming can also help to reduce water consumption, eliminate the use of harmful pesticides and fertilizers, and lower the risk of crop failure due to pests or disease. Vertical farming has the potential to create new opportunities for urban agriculture, making it possible to produce food in densely populated cities where traditional farming is not possible. The combination of lab-grown meat and vertical farming could fundamentally transform the food industry, disrupting the traditional model of farming and livestock production. If successful, these innovations could help to address some of the most significant challenges facing the food industry today, such as resource depletion, climate change, and public health concerns related to food safety and nutrition. There are still many questions that need to be answered before these innovations can become mainstream. First, there is the issue of scalability. Currently, lab-grown meat production is very expensive and requires extensive laboratory equipment and technology. It is unclear how this technology will be scaled up to meet the growing demand for meat while still being cost-effective. Similarly, vertical farming requires a significant investment in infrastructure and technology, which may limit its applicability to certain regions or markets. There is the issue of consumer acceptance. While some people may be excited about the prospect of lab-grown meat and vertical farming, others may be skeptical or even afraid. There may be a need for education and communication to inform consumers about the benefits and safety of these innovations. Another challenge that needs to be addressed is regulatory approval. Both lab-grown meat and vertical farming will require regulatory approval from governments around the world before they can be sold to consumers. This can be a lengthy and complex process, particularly given the innovative nature of these technologies. It is unclear how long this process will take or what the regulatory landscape will look like once these technologies are approved. There is the issue of cost. While lab-grown meat and vertical farming have the potential to reduce resource consumption and improve food safety and nutrition, they may also be more expensive than traditional methods of food production. This could limit their accessibility to certain segments of the population, particularly those with lower incomes or living in food-insecure regions. There may be a need for government subsidies, grants, or other forms of financial support to ensure that these innovations are accessible to everyone. Lab-grown meat and vertical farming are innovations in food production that have the potential to transform the way we eat. They offer several benefits, including improved sustainability, animal welfare, and food safety and nutrition. There are still many challenges that need to be addressed, including scalability, consumer acceptance, regulatory approval, and cost. It will be important to address these challenges in order to ensure that these innovations can be successfully integrated into the food industry and make a positive impact on our planet and our health.

VI. VERTICAL FARMING

Vertical farming is an increasingly popular practice in agriculture that involves growing crops in vertically stacked layers using advanced technological systems. The primary purpose of this approach is to optimize land use, prevent soil degradation, reduce water consumption, minimize transportation costs, and promote year-round crop production. Essentially, it aims to address some of the challenges facing conventional agriculture, such as limited land availability, unpredictable weather patterns, and pest infestations. There are several types of vertical farming systems, including hydroponics, aeroponics, and aquaponics, which involve cultivating crops in mineral-rich water solutions, air, and nutrient-rich water, respectively. These systems typically utilize artificial lighting, temperature, and humidity controls to create the ideal growing conditions for crops and maximize yields. Vertical farms can be located in urban areas, where fresh produce is typically scarce or can be transported over long distances, in remote locations with harsh climatic conditions, and in areas with low soil quality or high population density. The concept of vertical farming dates back to the early 20th century, but it was not until recently that the idea gained traction as a sustainable solution to global food security challenges. Currently, this approach is gaining popularity and is projected to continue growing in the coming years. The global vertical farming market was valued at $2.23 billion in 2018 and is expected to reach $12.77 billion by 2026, at a compound annual growth rate of 24.8%. The growth is driven by the rising demand for fresh and locally produced crops, increased investments in research and development, technological advancements, and the need for re-source-efficient and environmentally-friendly food production methods. The potential benefits of vertical farming are vast and far-reaching. One of the significant advantages of vertical farming is that it enables farmers to produce more food per unit of land compared to traditional agriculture. This is because the crops are stacked vertically, allowing more cultivation per square meter than in conventional farming, where large areas are needed to grow a limited variety of crops. Vertical farming systems use controlled environments that optimize the growth of crops, reducing the time they take to mature and increasing the yield per unit of area. Vertical farming systems can be designed to operate year-round, unlike traditional agriculture, which is highly dependent on seasonal changes. Another advantage of vertical farming is that it consumes significantly less water than is necessary for conventional agriculture. Vertical farming systems typically require only 70-90% less water than traditional farming methods, as the water used is recycled through the same system continually. The closed-loop systems enable farmers to use water efficiently, thereby reducing water runoff, soil erosion, and nutrient depletion. Vertical farming also minimizes the need for pesticides and herbicides, due to the controlled environment of the crops, which are less exposed to pests and diseases. This reduces the risk of contaminating the crops and the environment, promoting sustainable, and organic farming practices. Vertical farming also offers several economic and social benefits. Due to its location in urban areas and proximity to consumers, it reduces the transportation costs of food, reducing food waste and increasing the availability of fresh and locally grown produce. Vertical farms can create new job opportunities and improve the livelihoods of small-scale farmers, especially in urban areas where space is limited. Vertical farming can also serve as an educational and community resource, providing opportunities for students and residents to learn about sustainable and healthy food production and consumption. While vertical farming may seem like a panacea for all agricultural challenges, the technology has some limitations that need to be addressed. Firstly, the high capital costs associated with setting up a vertical farm may hinder its adoption by small-scale farmers and limit its accessibility to areas with low-income populations. Secondly, the technology does not address some of the root causes of food insecurity, such as unequal access to resources and power imbalance in the global food system. Proponents argue that vertical farming could serve as a complementary solution to the traditional agriculture system and increase food resilience in times of global food crisis. Vertical farming is a promising innovation in agriculture that has the potential to revolutionize how we produce and consume food. The technology has the potential to address some of the challenges facing traditional agriculture, such as limited land availability, unpredictable weather patterns, and pest infestations. Vertical farming could provide several benefits, such as increased yields per unit of land, reduced water usage, and minimal pesticide use, thereby promoting sustainable and organic farming practices. While the technology has some limitations, such as high capital costs, it provides a complementary solution to the global food insecurity challenge and should be explored and developed further to maximize its potential.

THE DEFINITION OF VERTICAL FARMING

Vertical farming is a modern agricultural technique that involves growing crops in stacked layers using artificial lighting, controlled temperature, and humidity to optimize plant growth. The practice aims at producing crops all year round in urban areas where land is scarce, and the climate is not favorable for crop growth. The farming practice is a solution to many problems associated with traditional farming, including inadequate space for food production, depletion of natural resources, and increased dependence on food imports from other regions. The technique is highly efficient in terms of space usage as vertical farms utilize a small fraction of land that traditional farms use to produce the same amount of crops. This is mainly achieved by stacking layers of crops, and as such, this approach has the potential of providing food to a rapidly growing population that is set to exceed ten billion by 2050. The vertical farming system employs the use of hydroponics to grow crops. Hydroponics is a soilless system that utilizes water and mineral nutrient solutions to feed the plants. This technology delivers nutrients to the roots of crops through a recirculating system that collects and reuses the nutrient solution. The farming technique is environmentally friendly since it conserves water, reduces the need for pesticides, avoids soil erosion, and reduces fertilizer use and runoff that can pollute water resources. This technology has made it possible to grow crops all year round with consistent yields, regardless of climatic elements such as temperature, rainfall, and sunlight, which have an impact on traditional farming. Vertical farming is not just about growing crops indoors, but it involves the use of advanced technology, such as artificial lighting, ventilation, and air conditioning to create optimal conditions for crop growth. The practice ensures that plants grow under the perfect conditions for their needs, resulting in high-quality plants that fetch higher market prices. Vertical farming techniques also increase the speed of crop growth by providing plants with the necessary conditions for faster growth and development. Vertical farms significantly reduce the cost of food production, transportation, and distribution since they are located in urban areas, which are closer to the market. The benefits of vertical farming extend beyond food production since it creates job opportunities in urban centers. The technique presents an opportunity for more people to engage in farming as it allows for the use of automation and robots for specific tasks such as planting, harvesting, watering, and nutrient delivery. This creates employment opportunities for urban dwellers who may lack traditional farming skills and knowledge. As the population continues to grow, so does the demand for food and additional land for agriculture. Vertical farming provides an innovative solution to this problem by maximizing land usage, reducing water wastage and minimizing the use of harmful pesticides. This method enables food production under any circumstances, even in places where conventional farming is limited by weather, space, or climate. The use of vertical farming techniques is also making food production more sustainable. Traditional farming is responsible for significant environmental problems, such as soil degradation, water contamination, deforestation, and greenhouse gas emissions. Vertical farming, on the other hand, greatly reduces the environmental impact of agriculture and ensures sustainable agriculture practices for future generations. The benefits of vertical farming extend beyond food production. It provides opportunities to develop new ways of cultivating crops and improving agricultural productivity and sustainability. The food industry is undergoing a revolution, and the use of vertical farming technology is one of the most significant developments that will shape the future of food production. There are still some challenges in the implementation of vertical farming. The vertical farming system requires a significant amount of energy to power the artificial lighting and temperature control systems. This can drive up the cost of production, and the resulting increase in food prices may discourage its widespread adoption. The technology required for vertical farming can be expensive, making it unavailable to small farmers. Vertical farming also requires skilled labor to manage the systems and maintain the equipment, which may also contribute to increased costs. Vertical farming is an innovative and cost-effective solution to the challenges associated with traditional farming. It provides an opportunity for food production in urban areas and creates new job opportunities. With the global population set to exceed ten billion in a few decades, the demand for sustainable food production systems is higher than ever. Vertical farming addresses this challenge by providing a new and sustainable way of cultivating crops, irrespective of location, climate, or weather. Despite the challenges, the potential benefits are massive, and adopting sustainable techniques like vertical farming can make food production more efficient and environmentally friendly. It is a promising technology that has a significant place in the future of food production.

THE BENEFITS OF VERTICAL FARMING COMPARED TO TRADITIONAL FARMING PRACTICES

Vertical farming is an innovative food production system that offers numerous benefits compared to traditional farming practices. One of the main advantages of vertical farming is that it is incredibly resource-efficient. Unlike traditional farming, which requires vast tracts of land, vertical farms can be constructed in relatively small spaces and can be located in urban areas, closer to the consumers they serve. This means that produce can be grown locally, reducing the need for long-distance transportation and the associated carbon emissions. Vertical farms can grow crops year-round, regardless of weather conditions, which means that they can produce more food per unit of land and water than traditional farms. This is particularly important in regions with scarce water resources, where vertical farms can use up to 90% less water than traditional farming practices. Vertical farms use fewer pesticides and herbicides than traditional farms, as they are grown in a controlled, sterile environment and are less susceptible to pests and diseases, which means that safer, healthier, and cleaner produce is produced. Because vertical farms are indoors, there is less pollution and runoff from fertilizers and other chemicals, leading to a cleaner environment. Another benefit of vertical farming is its scalability. Unlike traditional farming methods, where the amount of land available limits the amount of food that can be grown, vertical farms can be built high, enabling greater production per unit of land. This scalability allows for the production of more food with fewer resources and allows vertical farms to operate on a larger scale to meet the needs of growing populations. With a projected 70% of the world's population living in cities by 2050, vertical farming could be a vital component of a sustainable food production system. Vertical farming is an innovative, technology-driven approach to food production that has numerous advantages over traditional farming methods. Advances in lighting, irrigation, and automation technology allow vertical farms to maximize yields and minimize waste, making them more efficient than traditional farms. For example, nutrient solutions can be matched to the specific needs of each crop, and lighting can be adjusted to mimic natural sunlight conditions, helping to increase yields. Vertical farms can operate with a high degree of precision, allowing farmers to monitor and optimize crop growth in real-time, leading to consistent, high-quality produce and increased crop yields. Advancements in artificial intelligence and robotics could enable further automation of vertical farms, making them even more efficient and productive. For example, robotic systems could enable vertical farms to operate 24/7, leading to even greater yields and lower production costs. Vertical farming represents a significant step forward in the quest for more sustainable, efficient, and innovative food production systems. By leveraging cutting-edge technology and innovative approaches to food production, vertical farms offer numerous benefits over traditional farming practices, including increased resource efficiency, scalability, and precision. As a result, vertical farms are poised to play a critical role in the future of food production, particularly in urban areas, where space and resources are scarce. As the world's population continues to grow and as the demand for healthy, sustainable food increases, vertical farming has the potential to revolutionize the way we produce, distribute, and consume food, leading to a more sustainable, efficient, and equitable food system.

THE POTENTIAL DRAWBACKS OF VERTICAL FARMING

While the benefits of vertical farming are numerous, it also has its fair share of potential drawbacks. One primary concern is the high cost of setting up and maintaining a vertical farm, which might limit its accessibility to small-scale farmers. The initial construction of vertical farms may require a significant capital expenditure, including high-tech machinery and equipment, lighting systems, and indoor climate regulation. The high cost of energy for controlled environments may produce significant operational expenses. The electricity to power the lighting systems, the air conditioning systems, and the water pumps required to run the farm could prove quite costly. Consequently, for vertical farming to become an economically viable enterprise, it may require significant subsidies or investments by private investors or government agencies. This could have particular implications in economically challenged regions where such a significant capital investment may not be tenable. Another issue that could impede the widespread adoption of vertical farming is the limited crop variety that can be grown within this environment. While vertical farming can grow household staples such as lettuce, herbs, strawberries, and kale, it may not be possible to grow other crops, such as beans or carrots, that require more extensive root systems. These crops can grow too wide and deep for vertical farms' relatively confined, and controlled conditions, making it difficult for the plants to have the proper nutrients and space for proper growth. Although crops that have different nutrient requirements can find their place in stacked plants, it is difficult to know how different plants' growing needs might interact when they are grown close to each other. Since vertical farming requires a high-tech and highly controlled environment, it also has significant energy demands, a situation that could negatively impact the environment. The energy needed to power the lighting, climate control, and water pump systems for vertical farms could prove to be unsustainable in the long run. For instance, the cost of energy required to power vertical farms in regions with an already deficient electrical grid could lead to power shortages during peak periods, pushing up energy prices, and potentially leading to blackouts. What's more, vertical farming could incentivize policies focusing on high-efficiency agricultural technologies at the expense of conservation programs that are designed to preserve forests and other ecosystems. This approach may prove unsustainable for the planet if not undertaken within a comprehensive environmental policy that seeks to balance the roles of technology and conservation programs in food production. Another potential disadvantage of vertical farming is that it requires significant amounts of water. Since the crops grown in vertical farms are grown indoors, they require an artificial source of water, which makes them more reliant on the local water supplies compared to traditional farming. The amount of water a plant needs to grow healthily depends on several factors, such as its type, environmental conditions, and its phase of growth. If not managed carefully, water scarcity could lead to reduced crop yields, crop instability, and even plant losses. The water used in vertical farming often needs to be treated and processed, which can be prohibitively expensive. An examination of the potential drawbacks of vertical farming reveals that while the technology is capable of producing numerous benefits, it is also not without its challenges. The high cost of setting up and maintaining vertical farms, especially in economically challenged regions, limited crop variety, significant energy demands, and the potential for water scarcity, are just a few of the challenges that will need to be addressed if vertical farming is to become more than a concept and reach its full potential. Collaborative efforts are needed between private investors and government agencies to make vertical farming a reality, as well as to provide support and subsidies for small-scale farmers who might be interested in adopting the technology. While these challenges are significant, they can be surmounted, and the potential benefits of vertical farming for both the environment and human populations are undeniable. By confronting the drawbacks of vertical farming head-on, it is possible to create an agricultural system that will redefine how we produce food, comfortably feeding a growing global population in a sustainable and environmentally-friendly manner. Lab-grown meat and vertical farming are two innovative approaches to food production that are likely to revolutionize the way we eat in the future. By reducing the need for traditional livestock farming and large land areas, these technologies have the potential to address some of the biggest problems facing modern food systems, such as environmental degradation, animal welfare concerns, and food security. At the same time, they present new challenges and ethical dilemmas that need to be addressed before they can be fully embraced by consumers and policymakers. Lab-grown meat, also known as cultured meat or cell-based meat, is produced by cultivating animal cells in a lab and using them to create meat products that are virtually identical to conventionally raised meat. This technology has the potential to replace traditional livestock farming, which is a major contributor to greenhouse gas emissions, deforestation, and water pollution. By producing meat in a controlled environment, lab-grown meat can also reduce the risk of foodborne illness and improve animal welfare by eliminating the need to raise and slaughter animals. While the concept of lab-grown meat is not new, recent advances in cell culture techniques and bioreactor systems have made it more feasible and economically viable. Several startups and research organizations are currently working on scaling up production and bringing lab-grown meat to market, with the first commercial products expected to hit shelves in the next few years. Despite its potential benefits, lab-grown meat raises several ethical and cultural concerns. Some people argue that it is unnatural and goes against the idea of eating food that is grown and harvested in a traditional way. There are also concerns about the safety and long-term health impacts of consuming lab-grown meat, as well as about the potential loss of jobs and livelihoods in the traditional meat industry. The high cost of production and regulatory hurdles associated with lab-grown meat could limit its availability and affordability for consumers. Vertical farming, on the other hand, is a method of food production that involves growing crops in vertically stacked layers using artificial lighting and climate control systems. By utilizing urban spaces and reducing the need for large land areas, vertical farming has the potential to revolutionize the way we grow and distribute food, especially in areas where arable land is scarce or contaminated. Vertical farms can also reduce the environmental impact of agriculture by minimizing the use of water, pesticides, and fertilizers, and by reducing transport emissions. Vertical farming is still a relatively new and experimental technology, but several startups and research organizations are exploring its potential applications. While there are some early success stories, such as the vertical farms in Japan that produce lettuce for 200,000 people, the technology is not without its challenges. One of the main issues facing vertical farming is its high cost, which is due in part to the energy-intensive equipment required to maintain optimal growing conditions. Another challenge is the limited range of crops that can be grown effectively in vertical farms, which may restrict its ability to replace traditional farming methods. Despite these challenges, vertical farming is recognized as a promising solution for some of the key problems facing food systems. By reducing the need for large land areas, it can help to conserve natural habitats and protect biodiversity. By producing food closer to urban areas, it can also improve food security and reduce the environmental impact of transportation. Vertical farming can provide opportunities for local job creation and entrepreneurship. Lab-grown meat and vertical farming represent two innovative and disruptive approaches to food production that have the potential to address some of the biggest challenges facing modern food systems. While they present several challenges and ethical dilemmas, they offer a glimpse of a more sustainable, safe, and equitable food future. As these technologies progress and become more widely available, it will be important for policymakers, consumers, and industry leaders to ensure that they are implemented in a responsible and sustainable way that takes into account the needs and concerns of all stakeholders. Only then can we fully realize the potential of these transformative technologies and create a more prosperous and resilient food system for future generations.

VII. THE ENVIRONMENTAL IMPACT OF VERTICAL FARMING

Vertical farming has been hailed as a sustainable solution to traditional agriculture practices. Its environmental impact must also be taken into consideration. While vertical farming can reduce the carbon footprint associated with transporting food, it requires significant energy input for lighting, heating, and cooling systems. The energy used in vertical farming comes from fossil fuels, leading to greenhouse gas emissions. Vertical farming systems require a considerable amount of space and resources, such as water and nutrients. Concerns about the amount of energy used by vertical farming systems are not unfounded. Earlier attempts at indoor farming, such as greenhouses, have been criticized for their high energy usage. In comparison, vertical farms use less energy due to their efficient lighting, cooling, and heating systems. Nevertheless, the amount of energy required to power vertical farms is still significant. According to some estimates, the energy consumption of vertical farms is three times higher than that of greenhouses and ten times higher than outdoor farming. The use of artificial lighting in vertical farming is one of the main contributors to its energy consumption. LED lights are commonly used in vertical farms and are highly efficient, consuming less energy than other lighting systems. They still require a substantial amount of power to operate. The amount of energy needed for lighting can be reduced by using natural light, but this is limited by the availability of sunlight. Some crops require a specific type of light spectrum to grow efficiently, which means that natural light may not be sufficient for all crops. Besides, any lighting system generates heat, and cooling systems are required to maintain the optimal temperature for plant growth. These cooling systems also consume a significant amount of energy. The cooling and heating systems used in vertical farming must be carefully designed to minimize their environmental impact. High-temperature crops such as tomatoes and cucumbers require cooling systems to function optimally. These cooling systems need to be energy efficient, and the heat produced by them can be recycled in different areas of the vertical farm. Heating systems are necessary for the cultivation of crops in colder climates, such as Iceland and Canada. Heating systems can be powered by renewable energy sources, such as geothermal and solar energy. By using renewable energy sources for heating, vertical farms can help reduce their greenhouse gas emissions. The water requirements of vertical farming are also a concern since soil-free systems rely on nutrient-rich water to feed plants. Compared to traditional farming, vertical farms can use up to 70% less water to produce the same amount of produce. The water used in vertical farming must be carefully managed to prevent water waste. Some vertical farms use recirculating water systems that capture and filter water from plant transpiration and irrigation runoff. These systems reduce water waste significantly and can also reduce the amount of fertilizer required, further decreasing the environmental impact of vertical farming. Another aspect of vertical farming's environmental impact is the use of plastic in packaging and hydroponics systems. While hydroponic systems reduce the need for pesticides and herbicides, their use of plastic liners and substrates can create waste. Many vertical farms use recyclable or compostable materials in their packaging to reduce their environmental impact, and some even use biodegradable plastics. These materials are still not commonly used in the industry, and their implementation would require a significant shift in the supply chain. Vertical farming has the potential to be a sustainable solution to our agricultural system's numerous environmental problems. It is essential to recognize that vertical farming is not entirely environmentally friendly. Its energy consumption, dependence on scarce resources like water and nutrients, and reliance on packaging materials make it an innovation with unique environmental challenges. Nevertheless, the vast strides made in the field of vertical farming call for more attention to be given to addressing these challenges through innovative sustainable solutions. By continuing to improve its environmental profile, vertical farming can become an indispensable tool for feeding our growing global population while protecting our planet.

THE ROLE OF VERTICAL FARMING IN REDUCING GREENHOUSE GAS EMISSIONS

One of the most compelling innovations in food production is vertical farming, which has the potential to significantly reduce greenhouse gas emissions and address a range of environmental challenges. Vertical farms are indoor, multi-level agricultural facilities that use artificial lighting, climate control systems, and hydroponic or aeroponic farming techniques to grow crops in a highly controlled and efficient manner. By eliminating the need for land, soil, and pesticides, vertical farms can reduce the impact of agriculture on natural ecosystems and conserve water and energy resources. Vertical farms can be located closer to urban centers, reducing transportation emissions and increasing access to fresh, healthy produce. One of the keyways that vertical farming can reduce greenhouse gas emissions is by reducing the carbon footprint of food production. Traditional agriculture relies heavily on fossil fuels for equipment, transportation, and fertilizers, which are all major sources of greenhouse gas emissions. Vertical farms, on the other hand, can operate using renewable energy sources like solar or wind power, and can recycle water and nutrients to minimize waste and reduce environmental impact. Vertical farms can produce more food per unit of land than traditional farming methods, which can help alleviate pressure on deforestation and land conversion, further reducing the carbon footprint of food production. Another way that vertical farming can mitigate carbon emissions is by reducing food waste. According to the United Nations, approximately one-third of all food produced for human consumption is lost or wasted, which generates greenhouse gas emissions and contributes to global food insecurity. Vertical farms can help address this challenge by producing food in a more efficient and precise manner, minimizing spoilage and reducing the need for transportation over long distances. By producing fresh, healthy food locally and on demand, vertical farms can also help shift consumer behaviors towards smaller, more frequent purchases, reducing the likelihood of food waste at home. A third way that vertical farming can reduce greenhouse gas emissions is by enhancing carbon sequestration and mitigating the effects of climate change. By producing crops in a controlled environment, vertical farms can optimize growing conditions and accelerate plant growth, which can increase carbon uptake and sequestration. Vertical farming can also incorporate advanced technologies like sensors, data analytics, and machine learning to optimize resource utilization and minimize waste, further reducing the environmental impact of food production. Vertical farms can incorporate ecological design principles that enhance biodiversity, soil health, and ecosystem services, which can promote the resilience and adaptability of agricultural systems in the face of climate change.

One of the key challenges facing vertical farming is its energy consumption and carbon footprint, which can be high if facilities rely on non-renewable energy sources or inefficient building design. These challenges can be addressed through the adoption of sustainable building practices, such as passive solar design, efficient lighting, insulation, and ventilation systems, and the use of renewable energy sources like wind or solar power. Research and development efforts are ongoing to improve the efficiency and sustainability of vertical farming technologies, such as the use of more efficient LEDs or the development of closed-loop nutrient systems that reduce water and fertilizer waste.

Despite these challenges, vertical farming has the potential to revolutionize the way we produce food, reducing the environmental impact of agriculture and increasing access to healthy, sustainable food. In addition to its carbon mitigation potential, vertical farming can help address a range of other environmental challenges, such as water scarcity, habitat loss, and soil degradation. Vertical farming can promote greater food security and resilience by reducing dependence on long-distance transportation networks, by providing opportunities for locally-produced food, and by enabling the cultivation of food in areas with limited access to arable land. Vertical farming represents a promising innovation in food production that can help mitigate greenhouse gas emissions and reduce the environmental impact of agriculture. By offering a sustainable and efficient alternative to traditional farming methods, vertical farming can help address some of the most pressing challenges facing the global food system today, including climate change, food waste, and food insecurity. While there are still technical and economic challenges that need to be addressed in the scaling up of vertical farming methods, it is clear that this technology has the potential to transform the way we produce and consume food, ensuring a sustainable future for generations to come.

THE SUSTAINABILITY OF VERTICAL FARMING IN TERMS OF LAND AND WATER RESOURCES

Vertical farming has been seen as a solution to the problem of scarcity of agricultural land and water resources. Traditional agricultural practices require large amounts of land and irrigation to grow crops, which has led to deforestation, loss of biodiversity, soil degradation and water scarcity in many parts of the world. Vertical farming, on the other hand, requires minimal land and water resources, making it an attractive option for cities and regions that face these challenges. Vertical farming systems can use up to 95% less water than conventional farming methods, as the water is recycled within the system and there is no need for irrigation. The plants are also grown in a controlled environment, in which temperature, humidity, and light are optimized, which results in higher yields than with traditional field crops. As vertical farms can be located in urban areas, they have the potential to reduce transportation costs and carbon emissions associated with the transportation of fresh produce from rural areas to urban centers. Another advantage of vertical farming is that it allows for year-round cultivation, which increases the potential for crop diversification. The controlled environment of vertical farms also offers protection from pests and diseases, which reduces the need for pesticides and herbicides that can be harmful to the environment and human health. One of the major challenges facing the sustainability of vertical farming is the use of energy to power artificial lighting and climate control systems. The high energy consumption of vertical farming systems has been a significant barrier to their widespread adoption due to the high energy costs involved. While renewable energy sources such as solar panels can be used to offset some of the energy costs, it may not be enough to make these high-tech indoor farms economically viable in the long run. While vertical farming may be able to grow a significant portion of our food in controlled environments, it is not a complete solution to food security. The production of staple crops such as rice and wheat may still require large areas of land, and vertical farming may not be able to replace traditional farming entirely for crops that require soil and grow outdoors. The expensive equipment required for a vertical farming system may not be affordable for small-scale farmers or developing countries that lack the necessary infrastructure and resources. The production of food in vertical farms requires specialized knowledge and skills, which may limit the possibility of widespread adoption. The financial and technical constraints of vertical farming may mean that it is only viable for companies that can invest in expensive equipment and employ highly skilled staff, which could result in the concentration of power and wealth in the hands of a few large agribusinesses. This could have negative implications for small-scale farmers and rural communities who may be unable to compete with these powerful corporations. Vertical farming shows great potential for reducing the environmental impact of traditional agriculture while providing fresh produce in urban areas, but it is not a complete solution to food security. Although it can save water and land resources, it requires high energy costs, specialized knowledge, and equipment. There is a risk of concentrating wealth and power in a few large agribusinesses, which could have negative impacts on small-scale farmers and rural communities. Vertical farming should not be seen as a panacea for food security but rather as one of many options for sustainable agriculture in the future. It should be used in combination with traditional farming methods, agroforestry, and sustainable land management practices to achieve a balance between food production and environmental conservation. Policymakers should invest in research and development, as well as supporting the development of training programs and agribusiness models that can help small-scale farmers and rural communities to benefit from the potential of these innovative food production methods.

THE ETHICAL IMPLICATIONS OF VERTICAL FARMING

While the technological advancements of vertical farming hold considerable potential for future food production, we must also consider the related ethical and social implications. One of them is the potential for further industrialization of agriculture. As the growing global population continues to put pressure on our planet's resources, large food corporations have increasingly turned to vertical farming as a solution. While this may help to secure the world's food supply, it may also lead to a further concentration of power and control over the food system in the hands of a few companies. In a world where only a handful of corporations control the vast majority of the food supply, we will become increasingly dependent on these companies for our basic sustenance. Such power asymmetries could lead to reduced access to food for many, particularly those living in developing countries, and to the exploitation of farmers and workers along the supply chain. Another potential ethical concern with vertical farming is its high initial cost and potential to widen the inequality gap. The high cost of building and operating vertical farms could limit their use to large corporations who have the resources to invest in such technologies. This would limit opportunities for small farmers who may not have the financial resources to compete, further exacerbating rural poverty and inequality. Vertical farming technologies may be primarily used to grow high-value crops such as vegetables and fruits, as opposed to staple food crops such as grains and cereals. This creates a potential ethical dilemma as the poor, who rely on staple food crops, may be unable to access the fruits and vegetables grown in vertical farms due to their higher cost. Vertical farming's environmental impact also raises ethical concerns, particularly in the context of its water and energy use. These technologies require large amounts of electricity and water to operate, and their adoption may increase pressure on freshwater resources and contribute to climate change. Such effects could disproportionately affect low-income groups, who may already be struggling with the effects of climate change such as drought, floods, and food insecurity. As cities continue to expand, more urban farmers might turn to vertical agriculture to meet their food needs. This could lead to further gentrification and urban displacement, particularly in already stressed low-income communities. To avoid these unintended consequences, we must put in place mechanisms to ensure that the costs and benefits of vertical farming are distributed equitably across society. Vertical farming could, however, potentially provide a solution to the ethical and environmental issues connected to conventional agriculture. It could enable us to produce food closer to the point of consumption, reducing transportation emissions, reducing food waste, and enhancing food security by making fresh food available year-round. The artificial and controlled environment of vertical farms helps eliminate the use of pesticides and other harmful chemicals, making them safer and healthier. Vertical farms may provide employment opportunities for the low-income communities that are already grappling with unemployment challenges by creating jobs in urban areas. If implemented in a transparent and community-oriented manner, vertical farming could be a sustainable way to encourage local food production, create stable jobs, and enhance environmental protection.

Vertical farming offers an exciting opportunity for transforming food production and enhancing food security. It also presents a range of ethical dilemmas, including the potential for further monopolization of the food supply, uneven access to high-value produce, high energy and water use, and increased urban displacement in low-income neighborhoods. We must put in place systems to minimize these risks by promoting transparency, collaboration, and the equitable distribution of benefits and costs associated with vertical farming. Only then can we ensure that this technology enhances our food system's sustainability and security without negatively affecting society, particularly vulnerable and marginalized populations. The innovations in food production have been gaining the attention of many in recent years. The trends in sustainable agriculture and the growing interest in meat alternatives have driven the development of new technologies that have brought us to the forefront of the future of food: lab-grown meat and vertical farming. While these two may seem unrelated, they share a common goal – to address the looming food crisis that the world is facing. Lab-grown meat is a promising technology that could revolutionize the way we produce meat. It is a method of replicating real meat, without the need for rearing and slaughtering animals. A few companies have already successfully produced lab-grown meat, and it's quickly becoming an alternative that could replace traditional animal farming. The technology works by taking a small sample of animal cells and placing them in a nutrient-rich environment. Over time, the cells grow and multiply, ultimately forming muscle tissue, which can be harvested and processed into meat. Lab-grown meat has many potential benefits. Firstly, it would mean that animals wouldn't have to be reared and slaughtered for meat, addressing the ethical concerns raised by many animal rights activists. It would reduce greenhouse gas emissions and reduce the risk of diseases spreading from animals to humans.

Vertical farming, on the other hand, is a way of growing crops in a vertically stacked system, instead of traditional farming methods, where crops are grown on flat land. The technique is used to grow food in a controlled environment, utilizing hydroponics, aeroponics, and aquaponics systems. Vertical farms can be built in small spaces, and they can be stacked on top of each other, maximizing the use of limited space. Vertical farming allows for crops to be grown all year round, optimizing the growth process by providing the right nutrients, lighting, and climate at every growth stage. By growing crops in these controlled environments, the growth cycle is much shorter, and production yields are significantly higher as compared to traditional farming. Vertical farming reduces water consumption by over 70% as compared to traditional farming. Together, these two technologies hold immense potential for transforming the way we produce and consume food. With the global population expected to reach almost 10 billion by 2050, the challenge of producing enough food for everyone will become more daunting. Traditional agricultural methods will be stretched beyond their limits, leading to various environmental challenges. Lab-grown meat and vertical farming are the perfect solution for this problem. By replacing traditional animal farming with lab-grown meat, we could prevent the emission of greenhouse gases, reduce the spread of animal diseases, and eliminate the ethical concerns of animal rights activists. Similarly, vertical farming could play a significant role in addressing the issues of food scarcity, deforestation, and land use. With vertical farming, we could grow crops and vegetables in cities and urban areas, reducing the need for transporting food over long distances. The controlled environment would also mean that crops could grow faster, with higher yields, resulting in more food production per unit of land area. While the benefits of these technologies are clear, there are a few challenges that need to be addressed. For lab-grown meat, the technology is still very new, and the costs of production are incredibly high. The cost of producing one pound of lab-grown meat could be as high $2,400, which would make it economically unviable. Companies would need to develop more efficient ways of producing lab-grown meat, reducing production costs, for it to be commercially successful. There will be challenges in gaining regulatory approval for the production and sale of lab-grown meat, as it is a new technology that is yet to be fully understood by consumers. For vertical farming, the challenges are different. The cost of establishing vertical farms and their operational expenses are three times higher than traditional farming methods. Affordability and the availability of resources could become major challenges that would limit the expansion of this technology. The energy required for lighting and environmental control could become significant, and it must be sustainably sourced for the technology to be truly environmentally friendly. Lab-grown meat and vertical farming are game-changing technologies that have the potential to transform the food industry and address the looming food crisis. These technologies could provide solutions to the challenges of increasing agricultural productivity, providing enough food for the growing world population, and reducing the environmental impact of farming. With more research and development, these technologies could become more cost-effective, scalable, and widely available. The future of food lies in these innovations, and as consumers, it is our responsibility to support the research and development of these technologies for a better, sustainable future.

VIII. VERTICAL FARMING AND HUMAN HEALTH

One of the most significant advantages of vertical farming is that it produces crops without the use of harmful chemicals, such as pesticides and herbicides. Vertical farms use hydroponics, where plants are grown in nutrient-rich water, and aeroponics, where plants are grown in an air or mist environment and nutrient solution is sprayed over roots, eliminating the need for soil. These methods ensure the plants are protected from pests and disease, and the controlled environment makes it possible to optimize plant growth using minimal water and no synthetic fertilizers. This approach is particularly important in areas where soil health has declined, and traditional farming is becoming more challenging. Vertical farming can have a positive impact on human health. Studies have shown that exposure to pesticides, herbicides, and synthetic fertilizers can result in adverse health effects, including cancer, respiratory problems, neurological disorders, and reproductive and developmental issues. By using hydroponics and aeroponics to grow crops, vertical farming eliminates the need for these harmful chemicals, reducing the risk of exposure to millions of people worldwide. Vertical farming can provide fresh, nutrient-dense produce year-round, often grown closer to the consumer. As a result, it is a convenient option for urban dwellers who may not have access to fresh produce due to their location or affordability. Vertical farming can also address the global issue of food insecurity, particularly in urban areas. With the world's population projected to reach nine billion by 2050, traditional farming methods may not be enough to meet the growing demand for food. Established infrastructure and shifting demographics make it difficult to grow and transport global produce to urban areas. Vertical farming offers a sustainable solution by using limited resources and space to produce food locally. It eliminates transportation costs and reduces the carbon footprint associated with traditional farming methods. The continuous production of crops in a vertical farm means that harvests can occur more frequently than in traditional agriculture, ensuring a steady supply of fresh produce, which sustains food security. Vertical farming may also play a significant role in reducing the environmental impact of food production. The production of food is becoming increasingly dependent on fossil fuels, including petroleum-based fertilizers and transportation systems. Traditional farming practices have led to significant soil degradation, water pollution, and biodiversity loss. Vertical farms can help reduce these impacts by requiring less water and energy resources than traditional agricultural practices. Vertical farming can reduce the carbon footprint associated with food production by growing crops closer to the consumer and eliminating transportation costs, which are responsible for a significant percentage of greenhouse gas emissions. Despite the potential benefits of vertical farming, there are still some challenges that need to be addressed. One significant challenge is the cost of production, which is often higher than traditional agriculture due to the high energy costs and capital investments required to build and operate the infrastructure. Advances in technology and engineering are making it more affordable and efficient; as a result, vertical farming is becoming more accessible to communities worldwide. Another challenge is the limited range of crops that can be grown in a vertical farm; crops that require large amounts of space, such as wheat or corn, may not be feasible for vertical farming. Nonetheless, alternative crops like leafy greens, herbs, and berries can flourish in controlled vertical farming environments and provide valuable nutrition in urban areas. An issue associated with vertical farming is its reliance on artificial light. While the technology for using natural sunlight in vertical farms is still emerging, most vertical farms rely heavily on artificial lighting systems, which can be energy-intensive and lead to increased carbon emissions. Excessive lighting can cause sleep disorders, headaches, and eye strain, which raises concerns about the health of workers in vertical farms. Another challenge associated with vertical farming is the high-tech and potentially alienating nature of the facilities. Vertical farms are often large-scale, high-tech facilities, with little natural light or outdoor access. As such, they may feel less human-centered than traditional farming. There is potential to incorporate designs that make vertical farms more inviting and community-focused. For example, the Association for Vertical Farming has called for the use of vertical farms in urban areas to be integrated with schools, hospitals, and public spaces to promote healthy lifestyles and community engagement.

Vertical farming, while still a relatively new technology, has the potential to revolutionize our food systems and have a positive impact on human health, environmental sustainability, and food security. By eliminating the need for synthetic fertilizers and chemicals, reducing transportation costs and emissions, and enabling the localized production of food, vertical farming can provide fresh, nutritious produce year-round, regardless of location or weather conditions. Still, research and development are required to optimize the technology, improve energy efficiency, and minimize environmental impacts. Nonetheless, with continued innovation, vertical farming could provide one promising solution to the challenges of modern food production systems.

THE NUTRITIONAL VALUE OF VERTICAL FARMING PRODUCE

Vertical farming is a promising innovation in food production that has the potential to revolutionize the traditional agricultural industry. It is a method of growing crops in vertically stacked layers using a controlled environment. Vertical farming has many advantages over conventional farming, including its efficiency, sustainability, and the increased amount of food that it produces in a smaller area. One of the most important benefits of vertical farming is the high nutritional value of its produce. Vertical farms are able to create the perfect environment for crops to grow, which allows for optimal nutrient absorption. The nutritional value of produce grown in a vertical farm is unparalleled compared to traditional agriculture methods. Unlike conventional farming, vertical farms have the capability to control and manipulate the environmental factors that affect the growth and development of the plants. Vertical farms are able to provide the perfect amount of light, water, temperature, and nutrients that each crop needs. The ability to control these factors ensures that the plants grow faster, healthier, and with more nutrients than they would in a traditional farm. For instance, vertical farmers can provide the perfect mix of nutrients for each crop and adjust the pH level of the water to match the specific crop's needs. This precision means that the crops are able to absorb the nutrients more efficiently, resulting in a higher concentration of nutrients in the produce. The controlled environment in a vertical farm helps plants grow in a way that reduces the risk of

135

disease and contamination. The farms are able to eliminate the need for pesticides and herbicides, which are commonly used in traditional farming. This is due to the nature of vertical farming, which minimizes the chances of pests and diseases in crops, thereby reducing the use of herbicides and pesticides. Vertical farming is a closed system, which means that the plants are not exposed to external factors such as air pollution and toxins. This allows the plants to grow in a healthier environment, further enhancing their nutritional value. Another advantage of vertical farming is that it requires significantly less water than traditional farming methods. This is because the water in a vertical farm is recycled and reused, reducing the amount of water waste. The recycling of water also eliminates the need for pesticides and herbicides, which are known to contaminate water sources. In fact, vertical farms can reduce water usage by up to 70% compared to traditional farms. This reduction in water usage is reflected in the nutrient density of the produce. With less water in the crops, the nutrients in the plants are more concentrated, resulting in healthier and more nutritious produce. The controlled environment in vertical farms also allows for year-round production of crops. Vertical farms can produce crops every day of the year, regardless of external factors such as weather conditions and seasonality. This means that people always have access to fresh produce, which provides a consistent supply of nutrients in their diets. The year-round production also means that people can get essential nutrients that are usually not available during certain months, such as vitamin D from plants that are grown in an environment with UV light. Vertical farming also allows for the production of certain crops that cannot be grown in certain regions because of the environmental conditions. This provides people with access to different types of produce, ensuring that they get a balanced and varied diet. The nutritional value of vertical farming produce is enhanced by the fact that it is harvested and sold locally. Since vertical farms are often located in urban areas, it reduces the time between the harvest and the sale of the produce. This ensures that the produce is always fresh, which maximizes the nutrient content of the food. With traditional agriculture, the food often has to be transported to long distances, which often takes several days. In the process, the produce undergoes various processes, such as being coated with preservatives, which reduces its nutrient content. In traditional farming, crops are usually harvested before they are ripe as a result of the need to transport them long distances, which also reduces their nutrient content. In contrast, vertical farms can deliver freshly harvested produce directly to the market, eliminating this loss of nutrients. The nutritional value of produce grown in vertical farms is unparalleled compared to traditional agriculture. Vertical farms allow for optimal nutrient absorption by crops due to their ability to provide the perfect amount of light, water, temperature, and nutrients. The controlled environment of the farms helps to reduce the risk of disease and contamination, while the recycling of water reduces water usage. Year-round production and local harvesting ensure that the produce is fresh and nutritious. Vertical farming has the potential to revolutionize the way we produce food and address the global challenge of providing healthy food for a growing population.

THE POTENTIAL RISKS ASSOCIATED WITH VERTICAL FARMING PRODUCE CONSUMPTION

Despite the many benefits of vertical farming systems, there are potential risks associated with consuming produce grown in this manner. One of the main concerns is the accumulation of heavy metals and other contaminants in the plants. Heavy metals such as lead, cadmium, mercury, and arsenic can be present in soil and water and are toxic to humans even at low levels. Because vertical farming systems rely on nutrient-rich solutions rather than soil, there is a risk that these contaminants may accumulate in the plants if they are present in the nutrient solution. While some companies have developed systems that filter out heavy metals and other contaminants, these systems can be expensive and may not be accessible to all farmers. There is a risk that contaminants could enter the nutrient solution at various points in the production process, such as from the use of contaminated water or the presence of inappropriate fertilizers.

Another concern associated with vertical farming is the development of pesticide-resistant pests and diseases. Because vertical farms often operate in controlled environments, they may be more susceptible to outbreaks of pests and diseases that are resistant to conventional pesticides and fungicides. If left unchecked, these outbreaks could lead to serious yield losses and reduce the effectiveness of current pest management strategies. Some companies are using genetically modified organisms (GMOs) to increase crop yields and improve plant resistance to pests and diseases. While these GMO crops may be effective in the short term, there is concern about the long-term environmental and health impacts of these crops.

There are also potential risks associated with the use of artificial light in vertical farming systems. While artificial lighting can be used to achieve optimal growing conditions and maximize yields, it can also lead to excessive energy consumption and higher costs. There are concerns that prolonged exposure to artificial light could have negative effects on human health, such as disrupting circadian rhythms and increasing the risk of some types of cancer. To mitigate these risks, some vertical farming companies are exploring the use of natural lighting or energy-efficient LED lighting, which can help reduce energy consumption and minimize environmental impacts. Another potential risk of vertical farming is the use of hydroponic systems, which may not be as nutrient-rich as traditional soil-based systems. Hydroponic systems rely on nutrient-dense solutions to feed the plants, which can be difficult to balance and maintain. If not managed properly, these systems can lead to imbalances that reduce crop yields and damage plant health. Because hydroponic systems rely on chemicals rather than natural soil-based nutrients, there is a risk that consumers may not be getting the same level of nutrient-dense produce that they would get from traditional soil-based farming. There is the risk that vertical farming systems may not be as sustainable as initially thought. While vertical farms can help reduce water consumption and minimize the use of pesticides, fungicides, and fertilizers, there are still concerns about the energy-intensive nature of these systems. As mentioned, the use of artificial lighting can be energy-intensive and add to the overall carbon footprint of the farms. Because vertical farms typically rely on high-tech equipment and specialized facilities, the costs of building and operating these farms can be prohibitively high for many farmers. This may limit the accessibility of vertical farming to small-scale farmers, leading to the consolidation of the industry into larger, corporate entities.

Vertical farming systems offer many benefits to the food production industry, from maximizing yields and reducing water consumption to minimizing the use of pesticides and fungicides. There are potential risks associated with consuming produce grown in this manner, including the accumulation of heavy metals and other contaminants, the development of pesticide-resistant pests and diseases, and the use of hydroponic systems that may not be as nutrient-dense as traditional soil-based farming methods. The use of artificial lighting can lead to excessive energy consumption and higher costs, and the industry may not be as sustainable as initially thought. While efforts are being made to address these concerns, it is still important to carefully consider the potential risks associated with vertical farming before fully embracing this technology as a solution to feeding our growing population.

THE ROLE OF VERTICAL FARMING IN COMBATING FOOD-BORNE ILLNESS

In combating food-borne illnesses, one promising innovation is the practice of vertical farming. Vertical farming involves growing plants in vertically stacked layers, which maximizes the use of space and resources. Typically, the plants are grown in a controlled environment, such as a greenhouse or a building, using technology such as hydroponics and LED lighting. Vertical farming has several benefits when it comes to food safety. First, since the plants are grown in a controlled environment, the risk of contamination from external sources is minimized. This is especially important when growing leafy greens, which are particularly susceptible to contamination from pathogens such as E. coli and Salmonella. Second, since the plants are grown indoors, there is no need for pesticides or other chemicals that can contribute to food-borne illnesses. Instead, natural pest control methods, such as beneficial insects and microbes, can be used. Third, vertical farming allows for the production of fresh produce year-round, regardless of weather conditions. This means that produce does not have to be transported over long distances, reducing the risk of contamination from handling and storage. Vertical farming is a sustainable and efficient way to produce food, using only a fraction of the land and water required by traditional farming methods. This means that less land is exposed to potential contaminants, reducing the risk of food-borne illness. Since vertical farming can be done in urban areas, it reduces the distance that food has to travel to reach consumers, further reducing the risk of contamination. In addressing the problem of food-borne illness, it is important to consider the current state of the food industry. Despite advances in technology and food safety regulations, food-borne illnesses remain a significant problem in the United States. According to the Centers for Disease Control and Prevention (CDC), each year an estimated 48 million people get sick from a food-borne illness, 128,000 are hospitalized, and 3,000 die. This not only has a significant impact on public health but also on the economy. Food recalls and outbreaks can result in lost revenue for farmers and businesses, as well as increased healthcare costs and lost productivity. These outbreaks erode public trust in the food industry, which can have lasting consequences. While the food industry has made progress in reducing the risk of contamination, there is still room for improvement. This is where vertical farming can play a significant role. One of the key advantages of vertical farming is the ability to control the growing environment. This includes factors such as temperature, humidity, lighting, and nutrients. By optimizing these conditions, plants can grow faster and produce higher yields, while also reducing the risk of contamination from external sources. For example, by using hydroponics, plants can be grown without soil, reducing the risk of contamination from soil-borne pathogens. Since hydroponics recirculates water, it reduces the amount of water needed and the potential for contamination from runoff. Likewise, LED lighting allows for precise control over the amount and type of light that plants receive, optimizing growth and reducing the risk of pests and diseases. The use of natural pest control methods such as ladybugs and nematodes reduces the need for pesticides, further reducing the risk of chemical contamination. By controlling the growing environment, vertical farming can produce high-quality, fresh produce that is less susceptible to contamination, reducing the risk of food-borne illness for consumers. Another advantage of vertical farming is the ability to produce food year-round, regardless of weather conditions. This has several benefits when it comes to food safety. First, since the plants are grown indoors, there is no risk of contamination from environmental factors such as air pollution or animal feces. Second, since the produce is grown locally, it does not have to be transported over long distances, reducing the risk of contamination from handling and storage. Since the produce is grown year-round, there is no need for long-term storage, which can increase the risk of contamination. Instead, produce can be harvested and delivered to consumers quickly, ensuring maximum freshness and flavor. By reducing the distance that food has to travel and minimizing the risk of contamination, vertical farming can play an important role in improving food safety. The role of vertical farming in combating food-borne illness is significant. By controlling the growing environment and producing fresh, locally grown produce year-round, vertical farming can minimize the risk of contamination from external sources and reduce the need for pesticides and other chemicals. This not only benefits consumers but also farmers and businesses, who can reduce the risk of lost revenue from food recalls and outbreaks. Since vertical farming is a sustainable and efficient way to produce food, it has the potential to revolutionize the way we grow and consume food. As the population continues to grow and the demand for fresh produce increases, vertical farming will become increasingly important in ensuring that food remains safe, healthy, and available to all. Lab-grown meat and vertical farming are two of the most revolutionary innovations in the food industry today. Lab-grown meat has the potential to completely transform the way we produce meat, making it more sustainable, ethical, and environmentally friendly. Vertical farming, on the other hand, can revolutionize how we produce crops, allow us to grow food in urban areas, and reduce our reliance on traditional agriculture. Together, these innovations will have a significant impact on the way we eat and the future of the food industry. Lab-grown meat, also known as cultured meat or clean meat, is created by taking small samples of animal muscles and growing them in a lab. The process involves extracting stem cells from a live animal and placing them in a nutrient-rich environment where they divide and form muscle tissues. These tissues are then harvested and processed to create meat products. The technology to produce lab-grown meat has been around for some time, but recent advances have made it possible to produce cultured meat on a commercial scale. One of the main advantages of lab-grown meat is its potential to address many of the problems associated with traditional meat production. For instance, raising livestock for meat production is a major contributor to climate change, accounting for approximately 15% of global greenhouse gas emissions. Lab-grown meat, on the other hand, has a much smaller environmental footprint as it requires significantly less land, water, and other resources compared to traditional meat production. Cultured meat also has the potential to alleviate ethical concerns surrounding animal welfare. Many consumers are uncomfortable with the idea of animals being raised for slaughter, and lab-grown meat offers a cruelty-free alternative. The production of lab-grown meat is free from antibiotics, hormones, and other chemicals that are commonly used in animal agriculture, making it a healthier option. Despite these advantages, there are still some challenges to overcome before lab-grown meat can become a viable alternative to traditional meat production. The cost of producing cultured meat is still relatively high, and there are concerns about its safety and acceptance by consumers. The technology is still in its early stages, and there is much to be learned about how to optimize production processes and ensure the quality and taste of the product.

Vertical farming is another innovative approach to food production that has the potential to revolutionize the industry. In vertical farms, crops are grown in stacked layers, using artificial lighting, climate control, and hydroponic systems to create a controlled environment. This allows for year-round crop production in urban areas and reduces the need for traditional agriculture practices. One of the main advantages of vertical farming is its potential to address many of the challenges associated with traditional agriculture. For instance, traditional agriculture is very dependent on the availability of fertile land, which is becoming increasingly scarce due to urbanization and climate change. Vertical farming, on the other hand, can be set up in urban areas, allowing for crop production in areas where land is limited. Vertical farming can also significantly reduce water usage compared to traditional agriculture. It's estimated that traditional agriculture uses around 70% of the world's freshwater resources, making it a significant contributor to water scarcity. Vertical farming, on the other hand, uses a closed-loop system where water is recirculated and reused, reducing overall water usage. Vertical farming can reduce the need for pesticides and other chemicals typically used in traditional agriculture. The controlled environment in vertical farms allows for a more targeted approach to pest control, reducing the need for harmful chemicals. This makes vertical farming a more sustainable and eco-friendlier alternative to traditional agriculture. Despite these advantages, there are also some challenges to overcome before vertical farming can become a widely adopted practice. The cost of setting up a vertical farm is still quite high, and there are concerns about the energy usage required for lighting and climate control. There are still many questions about how to optimize crop production in vertical farms and how to ensure the safety and quality of the crops produced. Lab-grown meat and vertical farming are two of the most exciting innovations in the food industry today. Both have the potential to significantly impact the way we produce and consume food, addressing many of the challenges associated with traditional agriculture and meat production. Lab-grown meat offers a more sustainable, ethical, and environmentally friendly alternative to traditional meat production, while vertical farming allows for year-round crop production in urban areas, reducing our reliance on traditional agriculture. While there are still challenges to overcome, these innovations offer a glimpse into a more sustainable and eco-friendly future for the food industry.

IX. THE COMBINATION OF LAB-GROWN MEAT AND VERTICAL FARMING

As the world population continues to increase, the need for sustainable food production practices becomes more pressing. Fortunately, innovations in food production such as lab-grown meat and vertical farming offer promising solutions to address these challenges. Combining these two practices can be a game-changer in providing efficient and sustainable food production. Lab-grown meat has gained attention as a potential solution to address the environmental and ethical concerns associated with traditional animal farming. Lab-grown meat offers a way to produce meat without sacrificing animals and the environment's welfare. Unlike conventional meat production, which requires vast land, water, and energy resources, lab-grown meat uses significantly fewer resources, making it a more sustainable option. Vertical farming is an innovative agricultural practice that involves growing crops in vertically stacked layers, using artificial light and controlled climate conditions. Vertical farming allows for year-round crop production in urban areas, reducing transportation costs and minimizing the need for harmful pesticides and fertilizers. The practice offers a sustainable way of producing fresh produce, especially in highly populated areas, where space is limited. The combination of lab-grown meat and vertical farming could bring even more substantial benefits for food production. Firstly, the two practices can work hand in hand to solve the pressing issue of land scarcity. As global populations continue to increase, the amount of land available for farming becomes limited. With vertical farming, crops can be grown in vertically stacked layers, utilizing unused space in urban areas. By combining this with lab-grown meat, food can be produced in a smaller space requirement, reducing the pressure on the limited available land. As lab-grown meat production takes significantly less time, it will create more space for crop production in vertical farms, increasing the overall food production capacity. Secondly, the combination of lab-grown meat and vertical farming can reduce the environmental impact of traditional animal farming. Traditional animal farming is responsible for significant greenhouse gas emissions, deforestation, and water pollution. In contrast, lab-grown meat production requires significantly fewer resources, which leads to lower carbon emissions. Vertical farming also minimizes agricultural footprint, reducing carbon emissions further, as crops are grown in a controlled and water-efficient environment. By combining these practices, the environmental impact is greatly reduced, providing sustainable food production practices that benefit the environment and public health. Thirdly, the combination of these two practices has the potential to revolutionize the concept of food security. Food security is a significant challenge, particularly in developing regions where traditional agriculture methods may not be sustainable. Lab-grown meat and vertical farming offer a way to enhance food accessibility and affordability, providing quality and safe food to masses at the most reasonable price. Lab-grown meat can provide an alternative protein source to support traditional agriculture, especially in areas where raising livestock is not possible. Fourthly, the combination can provide an ethical solution for meat production. Lab-grown meat significantly addresses animal welfare concerns associated with traditional animal farming and meat consumption. It ensures that meat production doesn't become a luxury that is limited to the rich. By growing meat in a lab and in vertical farms, the production process is entirely controlled, minimizing mistakes, and ensuring that consumers receive healthy and safe meat. Combining lab-grown meat and vertical farming has the potential to create job opportunities within the food industry. The use of technology in farming has already seen significant advancement, creating opportunities in robotic engineering, agricultural logistics, greenhouse installation, and maintenance, among others. By adding lab-grown meat to the mix, job opportunities may extend to the field of tissue engineering, biotechnology, and food processing. Combining lab-grown meat and vertical farming can bring significant benefits to the food production industry and the environment. By utilizing urban spaces and technology in food production, these practices can significantly reduce the environmental impact of traditional animal farming, address ethical concerns associated with meat consumption, enhance food accessibility and affordability, and create job opportunities. While both practices are still in their nascent stages, the potential they hold for the future of food production makes them an exciting prospect for those looking for sustainable food solutions.

THE POTENTIAL BENEFITS OF COMBINING LAB-GROWN MEAT AND VERTICAL FARMING

Combining lab-grown meat and vertical farming presents a number of potential benefits for our food system. Vertical farming, with its space-efficient design and controlled environment, could be an ideal way to grow the plant-based materials necessary for lab-grown meat production. The two technologies could be housed in close proximity, allowing for a more streamlined production process that minimizes waste and resource usage. Combining lab-grown meat with vertical farming could help to address some of the key challenges facing the food industry, including environmental sustainability, animal welfare, and food security. One potential benefit of combining lab-grown meat and vertical farming is that it could significantly reduce the environmental impact of animal agriculture. Animal agriculture is a major contributor to greenhouse gas emissions and has been linked to a range of environmental issues, including deforestation, water pollution, and biodiversity loss. By producing meat in a laboratory setting, we could eliminate the need for large-scale livestock production and drastically reduce the environmental impact of meat consumption. By growing the plant-based ingredients for lab-grown meat in a controlled environment like vertical farming, we could further reduce our environmental footprint by minimizing water usage and eliminating the need for pesticides and other harmful chemicals. A second potential benefit of combining lab-grown meat and vertical farming is that it could improve animal welfare. Livestock raised for meat production are often subjected to harsh living conditions, including cramped living quarters, poor nutrition, and exposure to disease. By producing meat in a laboratory setting, we could eliminate the need to raise animals for food, reducing the suffering of billions of animals each year. As lab-grown meat production becomes more efficient and cost-effective, it could eventually become the norm in the meat industry, further reducing the demand for traditional meat production and improving animal welfare on a global scale. A third potential benefit of combining lab-grown meat and vertical farming is that it could address issues of food security. With a growing global population and changing dietary preferences in many parts of the world, there is increasing pressure on our food system to produce more food with fewer resources. Lab-grown meat and vertical farming both offer potential solutions to this challenge by enabling us to produce food more efficiently and sustainably. By growing plant-based materials for lab-grown meat in a controlled environment like vertical farming, we could produce food year-round, regardless of weather conditions or other external factors. Similarly, by producing meat in a laboratory setting, we could eliminate the need for large amounts of land and water typically required for conventional meat production, freeing up resources to be used for other purposes. A fourth potential benefit of combining lab-grown meat and vertical farming is that it could provide consumers with more choices when it comes to their food. As plant-based diets become increasingly popular and concerns about the environmental impact of meat consumption continue to grow, lab-grown meat offers an innovative solution that could appeal to a wide range of consumers. By growing the plant-based materials for lab-grown meat in a vertical farm, we could create a sustainable and locally-sourced alternative to traditional meat production. This could help to reduce the carbon emissions associated with transporting food long distances and support local agriculture in communities around the world. A final potential benefit of combining lab-grown meat and vertical farming is that it could encourage innovation and collaboration in the food industry. As these two technologies continue to develop, there will be opportunities for companies and researchers to work together to create new and innovative products that meet the evolving needs of consumers. By building a sustainable and efficient food system based on these technologies, we could create a more resilient and adaptable food system that is better equipped to handle the challenges of the future. Combining lab-grown meat and vertical farming presents a range of potential benefits for our food system. From reducing the environmental impact of animal agriculture to improving animal welfare, addressing issues of food security, providing consumers with more choices, and encouraging innovation and collaboration in the food industry, these two technologies offer a pathway to a more sustainable and resilient food system. As we continue to explore the potential of these technologies and expand their use around the world, it is clear that the future of food is bright and full of possibilities. By embracing new and innovative approaches to food production, we can create a better world for ourselves and for future generations.

THE POTENTIAL DRAWBACKS OF COMBINING LAB-GROWN MEAT AND VERTICAL FARMING

While combining lab-grown meat and vertical farming presents a promising solution to mitigating the issues of traditional livestock farming, there are still potential drawbacks that must be considered. One of the most significant concerns is the environmental impact of the energy required for lab-grown meat production and vertical farming. Although vertical farming can offset the carbon emissions of traditional farming, it still requires a significant amount of energy to operate, from lighting to ventilation systems. Similarly, lab-grown meat production requires a large amount of energy, particularly in the form of heat and electricity, to maintain the necessary conditions for cell growth. A study from Oxford University found that the emissions from lab-grown meat production could potentially be greater per kilogram than those from beef cattle if the energy used to produce the cells comes from non-renewable sources. While the use of renewable energy sources can reduce these emissions, it still may not be enough to completely mitigate the environmental impact of these technologies. Another potential drawback of combining lab-grown meat and vertical farming is the high cost of the technology. Vertical farming requires significant upfront investment in infrastructure and technology, and while lab-grown meat production is projected to become less expensive as the technology advances, it is still currently more expensive than traditional meat production. This high cost could limit access to these technologies to primarily higher-income consumers, leaving lower-income individuals reliant on traditional livestock products. Investment in these technologies may divert resources from other important areas of food production, such as sustainable agriculture and soil health. The social implications of combining lab-grown meat and vertical farming must also be considered. While these technologies have the potential to reduce animal welfare concerns, the creation of these novel food products raises ethical questions about the nature of food and its production. The processes of lab-grown meat and vertical farming are disconnected from the traditional methods of farming and raising animals, which may cause some people to question the integrity of these new foods. These technologies may further remove people from the source of their food, potentially exacerbating issues of food deserts and food insecurity. The use of these technologies could also result in job displacement in the traditional meat and dairy industries, highlighting the need for just transition policies to support workers through the transition to new food production methods. Another potential drawback of combining lab-grown meat and vertical farming is the loss of biodiversity. Traditional farming practices rely on a variety of crops and animals, each with unique genetic traits, to ensure resilience and adaptability to changes in the environment. By contrast, lab-grown meat and vertical farming rely on uniformity and consistency in their products, with little to no genetic diversity. This uniformity can increase the risk of disease and pest outbreaks, reducing the resilience of these production methods to environmental stressors. Similarly, monoculture farming, which is often associated with vertical farming, can lead to the depletion of soil nutrients and biodiversity loss, further contributing to environmental degradation. Combining lab-grown meat and vertical farming could have unforeseen consequences for human health, particularly in the long-term effects of consuming these novel foods. While the technology for lab-grown meat production is still in its early stages and not yet widely available, concerns have been raised about the use of growth hormones and antibiotics in the process of lab-grown meat production. Similarly, vertical farming may rely heavily on hydroponics, which uses nutrient solutions to grow plants and eliminates the need for soil. While this method can be effective in reducing pests and other issues associated with conventional farming, the long-term effects of consuming produce grown in this way are not yet fully understood. The introduction of novel food products may cause allergic reactions or other adverse health effects, further highlighting the need for careful testing and regulation. The combination of lab-grown meat and vertical farming presents a potential solution to many of the issues of traditional livestock farming and conventional agriculture. As with any new technology, there are potential drawbacks that must be carefully considered. These include the environmental impact of the energy required, the high costs of the technology, social and ethical implications, loss of biodiversity, and potential long-term health effects. As adoption of these technologies continues to grow, it is essential to address these challenges and ensure that these innovations do not create further harm to the environment, society, or human health. By doing so, we can continue to push towards a more sustainable and equitable food system for all.

THE ROLE OF COMBINING LAB-GROWN MEAT AND VERTICAL FARMING IN ENSURING FOOD SECURITY

The food security crisis is an increasingly urgent issue facing humanity. Given the projected increase in global population to over 9 billion by 2050, it is essential to develop innovative and sustainable solutions to meet the demand for food. Two technologies have emerged as potential solutions to the food security problem: lab-grown meat and vertical farming. Lab-grown meat and vertical farming are two innovative approaches in food production that hold immense potential for ensuring food security. Combining these two novel approaches can address several of the pressing issues surrounding food security, such as land and resource use, animal welfare, and environmental degradation. This has led to a growing interest in their implementation, which could help to reverse the negative consequences of traditional livestock farming while ensuring that people have access to nutritional and sustainable food sources. Lab-grown meat and vertical farming are two novel approaches to food production that present complementary benefits. Lab-grown meat is produced through in vitro cultivation of animal tissue in a lab setting. This technology provides a solution to the environmental impact and ethical issues of conventional livestock farming. In contrast, vertical farming refers to the cultivation of food crops in stacked layers, making it an efficient and sustainable process that maximizes the use of space and resources. By combining these two technologies, it is possible to produce protein-rich food without the negative consequences traditionally associated with meat production. Lab-grown meat and vertical farming integrated into a sustainable food system could reduce greenhouse gas emissions, lower food prices, and promote food sovereignty.

In addition to reducing environmental degradation, lab-grown meat and vertical farming have the potential to address growing food demand. With the projected increase in the global population, food security is becoming increasingly challenging. Traditional food production methods are seemingly inadequate in meeting the projected demand. Lab-grown meat and vertical farming address the shortage of food by offering unlimited possibilities to increase the food supply without relying on land availability. Vertical farming can provide consistent food supply all-year-round, eliminating the need for crop rotation, while lab-grown meat can offer a way to supplement conventional animal food production. This combination makes it possible to meet food demand without further straining the planet's resources.

Combining lab-grown meat and vertical farming would also lead to reduced costs of food production. Traditional meat production requires large amounts of water, land, and energy; this is costly and inefficient. As resources become scarce, their cost will inevitably rise, making food production more expensive. In contrast, lab-grown meat and vertical farming require significantly fewer resources. Lab-grown meat, for example, can be produced in a controlled environment that reduces the need for water and feed, while vertical farming can produce higher yields per square meter using fewer resources than conventional farming. Integrating the two technologies can lead to a significant reduction in production costs, translating into more affordable food prices for consumers. Incorporating lab-grown meat and vertical farming into food systems would also impact animal welfare positively. Traditional animal food production systems have often been marred by unhygienic and inhumane practices. Livestock animals are usually confined in small spaces, subjected to stress, and exposed to unsanitary conditions, which compromise their welfare. In contrast, lab-grown meat eliminates the need for animal confinement and slaughter, reducing the number of animals subjected to suffering. Vertical farming, on the other hand, eliminates animal handling altogether. By integrating the two technologies, it is possible to address the animal welfare concerns of traditional animal food production by making it more humane and sustainable. Combining lab-grown meat and vertical farming can also result in reduced environmental damage resulting from traditional animal food production. Research has shown that traditional methods of animal food production have significant environmental costs arising from deforestation, greenhouse gas emissions, and water pollution. These environmental costs have potentially disastrous consequences for the global ecosystem, including climate change and biodiversity loss. Lab-grown meat and vertical farming have the potential to reduce the impact of food production on the environment by lowering the carbon footprint of animal food production. Lab-grown meat production reduces greenhouse gas emissions, while vertical farming uses fewer resources than conventional food production, reducing deforestation. The technologies' integration could, therefore, result in a more sustainable approach to food production that is less environmentally damaging. Combining lab-grown meat and vertical farming technologies have immense potential in addressing pressing issues surrounding food security. Lab-grown meat and vertical farming offer new ways to address the environmental impact, animal welfare concerns, and food shortages associated with traditional animal food production. By integrating lab-grown meat and vertical farming, it is possible to produce a sustainable, affordable, and nutritious food source with the potential to revolutionize how we produce and consume food. While there may be challenges in implementing these technologies, with the right policies and investment, these approaches could change the way we eat, making food production sustainable and equitable for all. Concerns about the sustainability and ethics of traditional animal agriculture have led researchers to explore alternatives. One of the most promising of these is lab-grown meat, or cellular agriculture. This process involves taking muscle cells from an animal, such as a cow or chicken, and then growing these cells in a nutrient-rich culture medium in a bioreactor until they form muscle tissue. This tissue can then be harvested and processed into meat products. Lab-grown meat has the potential to revolutionize the way we produce and consume meat. It could eliminate the need for the vast amounts of land, water, and feed required by traditional agriculture, while significantly reducing greenhouse gas emissions and animal suffering. Lab-grown meat can be produced with greater precision and consistency than traditional meat, potentially reducing foodborne illness and allowing for more efficient distribution and supply chains. Another innovative solution to the challenges of traditional agriculture is vertical farming. This practice involves growing crops in vertically-stacked layers, often indoors using artificial lighting, temperature, and humidity control. The goal of vertical farming is to maximize the use of space and resources while minimizing energy consumption, water use, and pollution. It can also allow for the year-round production of fresh, locally-grown produce in urban areas, reducing the need for long-distance transportation and storage.

Together, lab-grown meat and vertical farming have the potential to create a more sustainable and ethical food system. They offer opportunities to reduce the environmental impact of food production while providing safe, nutritious, and affordable food to a growing global population. Both technologies face significant challenges and limitations that must be addressed in order to realize their full potential. One of the main challenges facing lab-grown meat is the high cost of production. Currently, the process is still relatively expensive, with a single serving of lab-grown meat costing hundreds or even thousands of dollars. This is chiefly due to the high cost of the culture medium required to grow the cells, as well as the energy and infrastructure needed to maintain bioreactors. Scaling up production could help to reduce costs, but this will require significant investment in research and development, as well as regulatory approval. Another issue with lab-grown meat is its acceptance by consumers. Many people are hesitant to eat meat that is not produced in the traditional way, and there are concerns that lab-grown meat may not taste or feel the same as traditional meat. There is a lack of regulation and clear labeling standards for lab-grown meat, which may make it difficult for consumers to know what they are buying or to trust its safety and sustainability. Vertical farming also faces several challenges. One of the main limitations is the high energy consumption required to power artificial lighting and environmental control systems. Vertical farms require significant investment in infrastructure, such as HVAC systems, irrigation systems, and plant-growing equipment. These factors can make it difficult for small-scale operators to start up and maintain profitable vertical farms. Another issue with vertical farming is the limited range of crops that can be grown effectively in this setting. Most vertical farms are designed to grow leafy greens, herbs, and some fruits, but they are not well-suited for crops that require large amounts of space or sunlight, such as grains or root vegetables. This limits the potential impact of vertical farming on overall food production. Despite these challenges, lab-grown meat and vertical farming hold great promise for the future of food production. By reducing the environmental impact of food production, increasing access to affordable and nutritious food, and promoting ethical and sustainable practices, these technologies could help to address some of the most pressing challenges facing the global food system. Realizing this potential will require further research and investment in these innovative approaches, as well as a commitment from governments, businesses, and consumers to support their development and adoption. If we can overcome these challenges and embrace these new solutions, we may be able to create a food system that is healthier, more equitable, and more sustainable for generations to come.

X. FUTURE IMPLICATIONS OF LAB-GROWN MEAT AND VERTICAL FARMING

The future implications of lab-grown meat and vertical farming are immense and varied. One of the most apparent changes that might occur is that food production will become more sustainable and efficient. Since these methods of production will require much less land and water, they will reduce strain on traditional farming ecosystems. With vertically farmed fruits and vegetables, there would also be reduced issues with soil degradation, pesticide use, and water wastage. These methods also have the potential to lower greenhouse gas emissions resulting from feed production as well as methane released by livestock. Thus, the overall ecological impact of food production could be lessened by these emerging technologies, resulting in a more sustainable and environmentally-friendly agricultural industry. Lab-grown meat specifically could also increase the availability of meat products. Meat produced from cultured cells is the product of fast-growing in-vitro meat culture technology. Lab-grown meat offers an alternative to traditional animal farming processes, which could particularly appeal to consumers who are environmentally conscious or have ethical objections to traditional meat production. In addition to tackling concerns about animal cruelty, lab-grown meat production could also reduce the spread of infectious diseases from meat. Several challenges are likely to arise with such changes to the food production industry. The decreased emphasis on traditional farming methods may have implications for rural communities that depend on farming for their livelihoods. Thus, governments and stakeholders would need to plan for this economic shift and consider ways to mitigate its impacts. Public perception will be vital to the success of lab-grown meat, as there might be a perception that it's an artificial or even 'unnatural' substitute for 'real' meat may deter some consumers from accepting it. As such, a considerable amount of education on the subject would be required to help people understand how lab-grown meat functions and its potential role in the food industry. Another potential issue is affordability. Even though lab-grown meat is now being produced on a limited scale, there are significant costs involved, which makes it expensive to produce. With the expected proliferation of vertical farms, it would be necessary to determine whether they can produce food at a price point that is competitive with traditional farming methods. Large-scale adoption and upscaling could take considerable time and resources so the technology will have to keep developing at a rapid pace to meet future demand and reduce costs in the long run. The future of food production innovations could also lead to a possible reduction in global food insecurity. Vertical farming is effective in addressing limited arable land, which is becoming a challenge in feeding the world's increasing population. The technology could be instrumental in enabling cities to produce their food locally, reducing the need for imported produce and minimizing food miles. That being said, vertical farming has limitations in economies of scale, and it is not yet clear whether the technology can scale to produce sufficient quantities of food to meet the dietary needs of the global population. It's evident that lab-grown meat has the potential to have wide-ranging impacts on the food industry. Generally, animal farming is resource-intensive and poses a threat to many endangered animal species as many consumers demand animal-based protein products. Lab-grown meat provides a sustainable way to produce meat while preserving the environment, which would appeal to such environmentally conscious groups. The lack of high land-use and absence of raising animals reduce the carbon footprint of the industry significantly. In comparison to traditional meat production, the increase in production of cultured meat will have less overall water usage and greenhouse gas emission, contributing to the reduction of the industry's climate impact. By reducing land usage, particularly for grazing and feed crops, there will be a decrease in the overall ecological footprint of meat production. Lab-grown meats could provide improved nutrition to consumers, particularly those who need extra protein in their diets. The nutritional value of traditional meats tends to be inconsistent across breeds and even within the same breed. Because lab-grown meat is made up of cells, it could be tailored to provide specific nutrient levels, such as higher iron content for those with anemia. This makes the meat easier to produce and the resulting products more standardized, consistent, and nutritionally-sound. Lab-grown meat and vertical farming may also have significant impacts on the ethical and nutritional components of food production. Animal welfare has long been a concern for many people, and the availability of alternatives could lead to a sudden fall in the number of animals required for food production. Because lab-grown meat is made from muscle cells, it is entirely free of antibiotics, hormones, and other additives given to traditional meat during its production cycle. As such, lab-grown meat is safer and healthier, which will attract many consumers.

While the implications of these innovations may be far-reaching, negotiations and successful implementation will be necessary to avoid any unintended and negative outcomes that may occur.

THEIR POTENTIAL TO REVOLUTIONIZE THE FOOD INDUSTRY

The potential for lab-grown meat and vertical farming technologies could revolutionize the entire food industry. Lab-grown meat, or otherwise known as cultured meat, is an innovative and environmentally friendly option to traditional meat production. This method is achieved through taking a small sample of animal tissue and utilizing it to produce meat in a lab. This process allows for reducing the need for animal agriculture and consequent pollution and damage to the environment. The benefits of lab-grown meat are substantial, including the potential to reduce greenhouse gas emissions, the need for water and land use, and the potential to create a sustainable food source. It is crucial to recognize the threats that traditional meat production poses to the environment and our welfare. The animal agriculture industry is responsible for a significant amount of carbon emissions, with estimates ranging from 14% to 50% of total global greenhouse gas emissions. The mass production of meat requires an enormous amount of water, land, and feed, which can lead to deforestation, water pollution, and air pollution. The introduction of lab-grown meat as a more sustainable meat source can have a profound impact on the environmental devastation caused by traditional animal agriculture. While animal agriculture may pose an environmental concern, it is also a growing concern for animal welfare. The conditions in which animals can exist can lead to an inhumane and cruel way of life. Lab-grown meat is one solution to animal welfare, sidestepping the need for animal agriculture altogether. Without the need to raise and slaughter animals for consumption, lab-grown meat eliminates animal suffering, a benefit that cannot be overlooked. Vertical farming is another innovation that has the power to transform the food industry. This technology would create the ability to cultivate crops more efficiently and free from many external influences that currently exist in traditional farming. Vertical farming is a method of growing crops by stacking them on top of one another in a controlled environment that has the necessary components to simulate the best possible growing conditions. The controlled environment could create ideal conditions for plant growth, minimizing the amount of water and fertilizer required. This model would revolutionize the food industry as it would enable the growth of fresh produce in urban environments, freeing up land currently designated for agriculture and enabling fresh produce to be grown closer to where it's consumed. As many of our cities continue to expand, the challenges for conventional crop farming become even more apparent. Vertical farming optimizes space and the use of resources, making it a very promising way to produce food in urban areas. This system also reduces transportation costs and the need to import crops from other places, reducing the carbon footprint of the industry ultimately by reducing the distance the food must travel. This would also have a significant impact on food security by creating an alternative to traditional farming that requires an enormous amount of space and resources. Vertical farming can also be used as a way to grow crops in areas with extreme climates or regions with land that is typically unsuited for traditional farming. With vertical farming, land-intensive crops could be grown in a fraction of the amount of space usually needed. Scientists predict that vertical farming will increase agricultural productivity, leading to more significant yields of crops while reducing growing times and allowing for year-round cultivation. As the population continues to grow, the ability to produce more food using the same amount of space and resources is critical. Vertical farming methods would create this possibility by making use of underutilized urban space and optimizing the use of resources to generate a considerable amount of high-quality food while reducing unnecessary waste. The advent of lab-grown meat and vertical farming has the potential to revolutionize the food industry as we know it. Lab-grown meat, with its substantial environmental and animal welfare benefits, can be a game-changer for an industry that continues to contribute to the planet's damage. The ability of vertical farming to produce food in urban environments more efficiently and reduce the distance between farm and table can also have a powerful effect on the food industry. Not only will it provide fresh, high-quality food, but it can also have a significant impact on the environment, food security, and reduce waste associated with crop transportation. While both lab-grown meat and vertical farming technologies are still in their infancy, they are promising steps towards providing the world with a sustainable and abundant food supply. The world is facing significant challenges regarding population growth, environmental concerns, and food security. The products of these two innovations can help solve these pressing issues.

THEIR POTENTIAL TO COMBAT ENVIRONMENTAL AND HEALTH ISSUES

The potential of lab-grown meat and vertical farming to address pressing environmental and health issues is one of the most compelling reasons to invest in these innovative food production technologies. One of the most pressing environmental challenges facing the world today is climate change. The production of beef and other animal products is a significant contributor to greenhouse gas emissions, deforestation, and water pollution. In contrast, lab-grown meat has been touted as a sustainable alternative to traditional meat production because it requires fewer resources and generates fewer emissions. According to a 2011 study conducted by a team of researchers at the University of Oxford, lab-grown meat could reduce greenhouse gas emissions by up to 96% compared to traditional beef production, while requiring 45% less energy, 99% less land, and 96% less water. Lab-grown meat could be one of the most effective means of combating climate change and preserving natural resources by reducing agricultural emissions, deforestation, and water usage.

Another environmental challenge that lab-grown meat can help address is the depletion of ocean and land resources due to overfishing and hunting. The rise in demand for meat has led to overfishing of oceans, which is now threatening the ecosystem. Lab-grown meat could potentially reduce reliance on conventional animal farming and dependence on ocean resources. Vertical farming could help address a multitude of environmental problems. As the human population continues to grow, the demand for food and resources is on the rise, leading to unsustainable farming practices and deforestation. Vertical farming could mitigate such challenges by growing crops in vertical spaces using hydroponics and LED lighting, allowing for high-density crop yields in urban areas. This technology could substantially reduce transportation costs and decrease the carbon footprint associated with conventional agriculture. Lab-grown meat and vertical farming could also have significant positive impacts on public health. Meat consumption, particularly red and processed meat, has been linked to a range of illnesses, including heart disease, cancer, and type-2 diabetes. The process of lab-grown meat production allows for the control of the nutritional value of the meat by altering its fatty acid composition, protein level, and vitamin content. Lab-grown meats could provide the benefits of meat consumption without the associated health risks. Vertical farming allows the production of organic and non-GMO crops that are free from pesticides and other harmful chemicals, which have been linked to cancer and other health issues. The control of the environment in vertical farming allows the production of crops with shorter cultivation cycles, which means higher yields in shorter periods, ensuring that consumers have access to healthy, fresh, and nutritious produce throughout the year. When growing in a controlled indoor environment, the crops are less subject to pests, diseases, and harsh weather, which means a reduction in the use of synthetic pesticides, herbicides, and fungicides, which can accumulate in the soil and groundwater, leading to potentially adverse environmental consequences. The potential economic benefits of these innovations cannot be overstated. Vertical farming and lab-grown meat could provide new and valuable opportunities in the form of high-tech job creation and economic development. Lab-grown beef, for example, has generated significant media attention and attracted investment from venture capitalists and big food companies alike. Such investments also attract R&D funding, leading to new technological innovations and advancements in food production. This economic transformation can have significant industry-altering implications, from the way we source food ingredients to how we regulate and consume food altogether. Localized vertical farming could reduce the reliance on imports by enabling domestic food production, leading to more food security and control over pricing. It is clear that innovations in food production, such as lab-grown meat and vertical farming, have significant potential to address the pressing environmental, public health, and economic issues that our world is currently facing. While these technologies may not provide complete solutions, they can certainly be part of a larger strategy to achieve a more sustainable, equitable, and inclusive food system. They cannot be implemented on their own but must be supported by supportive policies, regulations, and stakeholder engagement. Only by working together can we ensure that these innovations reach their full potential and benefit humanity in the long run.

THEIR POTENTIAL TO PROMOTE SOCIAL JUSTICE IN FOOD SECURITY

In addition to the economic and environmental benefits discussed earlier, innovations in food production such as lab-grown meat and vertical farming also hold tremendous potential to promote social justice in food security. Firstly, lab-grown meat has the potential to address the worldwide issue of food scarcity, especially in developing countries. With the growing demand for meat, the current state of animal agriculture is highly unsustainable, and the majority of the world's population cannot access or afford meat as a source of protein. According to a report by the United Nations Food and Agriculture Organization, around 795 million people are undernourished globally, with the majority residing in low-income countries. Lab-grown meat has the potential to address this problem by providing a sustainable and affordable source of protein that can be produced and distributed more efficiently than animal meat. In this way, it can help to create a more inclusive and equitable food system.

Vertical farming, on the other hand, has the potential to address issues of food accessibility and affordability in urban areas. According to the United Nations, more than half the world's population currently lives in urban areas, a trend that is projected to increase in the coming decades. This rapid urbanization has led to increasing food insecurity in cities, particularly among low-income communities. Vertical farming, which can be set up in urban areas, has the potential to provide fresh produce at an affordable cost, thereby improving the health and well-being of urban populations. Vertical farms can be established in food deserts, areas lacking in healthy food options, thereby addressing the issue of food inaccessibility in such areas. Both lab-grown meat and vertical farming hold great potential in promoting sustainable and ethical food production practices. Animal agriculture is a significant contributor to greenhouse gas emissions and climate change, and it raises ethical concerns regarding animal welfare. Consequently, lab-grown meat provides a more sustainable and ethical alternative to animal meat, which can promote an environmentally conscious society. Vertical farms, on the other hand, employ a closed-loop system for water and nutrient management, and they use significantly less land and water than traditional farming methods. They reduce the need for transportation, thereby minimizing carbon emissions associated with food distribution. In this way, vertical farming can promote sustainable and eco-conscious food production practices.

Despite the potential benefits of lab-grown meat and vertical farming, there remain concerns regarding their social implications. In particular, the high cost of producing lab-grown meat and setting up vertical farms may pose a challenge to their adoption in low-income communities that need them the most. The products of these innovations may be perceived as an elitist food item, catering only to the wealthy and privileged. Policymakers must ensure that these innovations are accessible and affordable to all segments of society, regardless of socioeconomic status. These innovations may also have unintended consequences on traditional agricultural communities, particularly those that depend on animal agriculture. Lab-grown meat has the potential to disrupt the current agricultural sector by reducing demand for animal meat. Vertical farming can also threaten traditional farming methods and displace farmers who are not equipped to transition to the new farming system. Policymakers must ensure that traditional farming communities are provided with the necessary support and resources to transition to the new food production methods. To conclude, innovations in food production such as lab-grown meat and vertical farming hold tremendous potential in promoting social justice in food security. They address issues such as food scarcity, accessibility, affordability, and sustainability and promote a more equitable and inclusive food system. There is a need to ensure that these innovations do not result in unintended consequences that threaten traditional agricultural communities and exacerbate existing socioeconomic inequities. Policymakers must work towards creating a more accessible and equitable food system by prioritizing the needs of low-income communities and providing the necessary support for a just transition to the new food production methods. Lab-grown meat and vertical farming are two of the most innovative solutions to some of the challenges faced by the food industry. These developments represent significant improvements in terms of sustainability, ethics, and health. By providing a more sustainable source of meat and vegetables, they reduce the environmental impact of traditional agriculture, as well as the ethical issues associated with animal welfare. The economic benefits of these technologies arise from their potential to increase production and profits. Adapting these innovations comes with its challenges, and while it may take some time for these to become mainstream, it is inevitable that they will change the way we eat. Lab-grown meat involves producing meat artificially, without raising and slaughtering animals. This innovation involves taking a small sample of animal muscle tissue and using this as a seed to produce more tissue in a laboratory setting. The end result is a product that looks and tastes like meat, but that has been produced in a more sustainable and ethical way. Compared to traditional meat production, lab-grown meat has the potential to reduce greenhouse gas emissions, land use, and water consumption. This is because it does not require the use of antibiotics or feed, and the production process is significantly more environmentally friendly.

Another innovation is vertical farming, which involves growing plants in stacked layers, indoors. By growing plants in this way, vertical farms can use a fraction of the space required by traditional farming, and also eliminate the need for pesticides, herbicides, and other harmful chemicals. They also provide a more controlled environment to grow crops in, allowing for more precise control over factors such as temperature, light, and humidity. This not only results in higher crop yields but also eliminates the dependence on weather conditions, making it possible to grow crops all year round. The impact that these technologies will have on the way we eat is significant. For example, with the introduction of lab-grown meat, it will be possible to enjoy meat without the negative environmental and ethical impact that comes with traditional farming. Consumers can enjoy meat while knowing that it was produced without the need for animal slaughter, making lab-grown meat a more ethical choice. The reduction in greenhouse gas emissions and land use associated with this technology would have a positive impact on the environment. It's also plausible that charging consumers extra for meat that is considered morally or environmentally conscious could create another source of revenue for traditional meat companies. Similarly, the introduction of vertical farming can also revolutionize the way we access our food. By growing crops in urban environments, it is possible to reduce the distance that food travels from farm to table. This is important because it reduces the transportation emissions associated with traditional agriculture. Since vertical farming can be done all year round, regardless of weather conditions, it can provide a more reliable source of fresh vegetables to consumers, while eliminating the need for chemical fertilizers, herbicides, and pesticides. This bodes well for areas where geography or weather makes traditional agriculture difficult, making urban farming a viable alternative. The road to mainstream convergence is not free of challenges. While several companies are already producing lab-grown meat, research and innovation are still required. One of the main challenges is to produce the product on a large scale while keeping the costs low. Currently, lab-grown meat is still very expensive, and many people will not be able to afford it. Promotions and strategically placed marketing would be a key factor in easing such challenges. Another challenge is to gain public acceptance of the product, particularly among those who are skeptical about artificial meat. This will require consumers to have confidence in the safety of the product, as well as its taste and nutritional value. Similarly, the success of vertical farming also depends on the current status on consumer sentiment. While the notion itself is a novelty and surely has its benefits, there are certain questions that need to be answered. For example, will this really reduce the environmental impact on the planet? Does it have a taste that is on par with that produced by traditional farming methods? Are the costs of setting up a vertical farm enough to offset the savings in resources? What about the affordability of these vegetables? To address these questions and concerns, further research needs to be conducted to ensure that vertical farming has the potential to become a viable alternative for people who want to produce their own food or have ready access to fresh vegetables. Lab-grown meat and vertical farming represent significant and innovative solutions that can address some of the challenges that we face in our food systems. While there are still challenges to overcome, these innovations will no doubt change the way we eat, as well as the impact that we have on the environment and animal welfare. The future of food is in the hands of innovators and pioneers, and it will be exciting to see how these technologies will evolve, as well as how they will be adopted in the coming years.

XI. THE ROLE OF POLICY IN THE FUTURE OF FOOD PRODUCTION

As lab-grown meat and vertical farming have the potential to revolutionize food production, it is imperative that policy supports these innovations. The current food industry is riddled with issues ranging from environmental degradation to animal cruelty, which is why these alternative options seem all the more appealing. Policy has the power to either nurture or suppress these technologies, making it essential that lawmakers take a proactive role in shaping the future of food. The adoption of lab-grown meat, for instance, will require significant regulatory changes. Currently, the U.S. Department of Agriculture (USDA) oversees animal slaughter and inspection, meaning that it would need to expand its regulatory authority to include lab-grown meat. Similarly, vertical farming will need to navigate zoning regulations and local ordinances regarding land-use, water, and energy use. These are potentially significant roadblocks that can hinder the growth and spread of these technologies, and so policy efforts must be focused on addressing these hurdles.

Another reason policy has a critical role in shaping the future of food production is that it can address issues of accessibility and sustainability. Lab-grown meat, for instance, is projected to cost significantly more than conventionally produced meat, which could limit its access to lower-income consumers. To help alleviate this issue, policy must be implemented to ensure that lab-grown meat is affordable and accessible to everyone. This may mean subsidies, tax breaks, or even direct price controls to ensure that this innovation does not become a luxury item for only the wealthy. Vertical farming has the potential to reduce the carbon footprint of food production significantly. Policymakers will need to ensure that these farms are powered by renewable energy sources. This can be achieved through tax incentives for green startups and other policies to encourage the development of sustainable energy infrastructure. Policy can also play a crucial role in ensuring that lab-grown meat and vertical farming are held to high standards of safety and transparency. As these technologies are relatively new, there may be concerns regarding their potential health risks or the quality of their products. To prevent such issues, policymakers must create rigorous testing and inspection procedures that ensure the safety and quality of lab-grown meat and vertical farming produce. Transparency will be critical in establishing public trust in these innovations. Clear labeling and disclosure standards will be necessary to ensure that consumers understand the methods used in producing their food. These standards must be rigorously enforced to prevent deceptive marketing and to maintain public confidence in these technologies. Lastly, policy can also help address the ethical concerns surrounding food production. The current industrial model of agriculture relies heavily on the exploitation of animals, which is a source of controversy among animal rights activists. Lab-grown meat presents an opportunity to eliminate this issue entirely by providing animal-free meat. To achieve this, policymakers must ensure that the rights of animals are protected during the process of developing these products. Regulation must be implemented to govern animal use in research and testing, and companies must be held accountable if any unethical practices or abuses are found to be taking place. Vertical farming has the potential to remove the need for animal fertilizers and pesticides. This can significantly reduce the harm caused to animals by the current industrial model of agriculture and create a more ethical and sustainable food system. Policy has a crucial role in shaping the future of food production. The adoption of lab-grown meat and vertical farming technologies will require significant regulatory changes to overcome issues of accessibility, sustainability, safety, transparency, and ethics. Consequently, policymakers must take a proactive role in creating policies that support the development of these innovations while addressing the potential challenges they present. By doing so, we can create a food system that is healthier, more ethical, and more sustainable for all.

THE NEED FOR POLICY FRAMEWORKS TO REGULATE THE USE OF LAB-GROWN MEAT AND VERTICAL FARMING

As lab-grown meat and vertical farming begin to gain more traction in the food industry, it is critical that policy frameworks are in place to regulate their use. Lab-grown meat, also known as cellular agriculture, is created by taking animal cells and growing them in a controlled environment, thereby bypassing the need for traditional animal slaughter in the meat industry. Meanwhile, vertical farming involves the cultivation of crops in controlled environments using artificial lighting and nutrient-rich liquid instead of soil. While both of these methods have the potential to revolutionize the food industry, they also come with unique sets of challenges that need to be addressed through policy frameworks. One of the key areas where policy frameworks are needed is in regulating the safety and health aspects of lab-grown meat. As this is a relatively new technology, there is still much research needed to determine the long-term health impact of consuming lab-grown meat. There are concerns around the potential for contamination during the creation process, given the risks associated with culturing animal cells. Policies around the production, processing, storage, labeling, and distribution of lab-grown meat will need to be set to ensure its safety for consumption. Lab-grown meat is still subject to the public's perception of what counts as "real meat." The question of whether lab-grown meat should be considered meat at all is still up for debate, with animal welfare groups arguing that it should be classified as a "food product" instead of meat. This distinction can have implications for how lab-grown meat is regulated, marketed, and sold. As such, policymakers will need to take into account public perception and opinion when setting policies around the lab-grown meat industry. Vertical farming also has significant potential to disrupt traditional food production methods. The controlled environments of vertical farms can create ideal growing conditions that can enhance the productivity of crops while reducing the use of pesticides and herbicides. It can minimize the impact of climate change on the agricultural industry by reducing water usage and using renewable energy sources to power indoor farms. There are also several challenges that need to be addressed through policy frameworks.

One of the challenges associated with vertical farming is the intensive use of resources such as water and electricity to create the ideal growing conditions. While using sustainable energy sources can mitigate the environmental impact, policymakers need to ensure that vertical farms are not misusing resources or creating adverse impacts on the surrounding environment. There also needs to be transparency around the sourcing of the crops and the use of fertilizers to ensure that the food produced in vertical farms is both environmentally sustainable and safe for consumption. Another challenge associated with vertical farming is the potential for it to displace traditional agricultural jobs, particularly in rural areas. Policies need to be set to compensate for the loss of jobs and ensure that farmers and agricultural workers are not left behind in the transition to vertical farming. This could include establishing worker retraining programs and offering financial support for farmers to make the transition to vertical farming. Despite the challenges, both lab-grown meat and vertical farming offer significant potential to revolutionize the food industry, particularly in terms of sustainability and efficiency. Lab-grown meat has the potential to eliminate animal slaughter and reduce the environmental footprint of the meat industry, while vertical farming can create sustainable food production methods that are resilient to climate change. Both technologies can improve food security and reduce reliance on imports. To fully realize the potential of these technologies, policymakers need to take a proactive role in creating regulatory frameworks that support their development and implementation. Policies need to be established around safety, health, labeling, and environmental sustainability to ensure that these technologies are used to their full potential without compromising public safety or the environment. Policymakers need to ensure that traditional farmers and agricultural workers are not left behind in the transition to these new technologies. The development of regulatory frameworks is essential to ensuring the success of lab-grown meat and vertical farming. These technologies have the potential to revolutionize food production and address some of the most pressing issues facing the food industry, including animal welfare, sustainability, and food security. They also come with unique sets of challenges that need to be addressed through regulatory frameworks. By developing policies that prioritize safety, health, environmental sustainability, and social equity, policymakers can support the development of these technologies while ensuring that they have benefits for everyone.

THE POTENTIAL ROLE OF POLICIES IN PROMOTING SUSTAINABLE AND EQUITABLE FOOD SYSTEMS

While lab-grown meat and vertical farming offer promising innovations for food production, their potential impact on the future of food will be determined by the policies that govern their implementation and operation. Policymakers can play a critical role in promoting sustainable and equitable food systems by crafting policies that ensure these innovations are used to advance social, economic, and environmental goals. Firstly, policymakers can promote sustainable food systems by incentivizing the adoption of eco-friendly farming practices and reducing food waste. Traditional industrial farming is often associated with the overuse of fertilizers, depletion of soil health, and pollution of water resources. With intensive vertical farming technologies, which often involve hydroponic systems, adequate amounts of water and nutrients can be provided to plants grown within air-controlled environments. Policymakers should encourage the promotion of sustainable practices such as the use of renewable energy sources like solar and wind power in these vertical farms. Implementing policies that reduce food waste can have a significant impact on the environment and meet the food needs of an increasing global population. By supporting the reduction of food loss and waste through policy, food production can be improved while also reducing greenhouse gas emissions. Secondly, policymakers can promote equitable food systems by ensuring access to nutritious and affordable food for all. Traditional farming practices often perpetuate economic disparities, particularly in rural areas where smaller farmers are outcompeted in a system that prioritizes large-scale farming methods. The use of vertical farms can create new food production opportunities that can challenge the status quo of traditional food production. Incentivizing small-scale vertical farming can enable greater economic benefits for smaller rural farmers while providing local communities with a source of affordable, healthy food. Municipalities can support the establishment of vertical farms in urban areas, providing residents with fresh produce. This model is especially important in low-income residential areas, where communities have limited access to fresh, high-quality produce.

Thirdly, effective policies can promote technological innovation to drive food production and distribution advances. It should not be forgotten that the vertical farming industry, in particular, is still in its early stages of development, and policymakers must ensure that it continues to thrive. By supporting research and development in advanced food production technology, policymakers can help develop the necessary infrastructure and practices required for industry growth. Policies that incentivize the use of advanced technologies such as artificial intelligence, robotics, and data analytics can help to reduce waste, improve production efficiency, increase food security and sustainability, and promote the deployment of new farming methods.

To conclude, lab-grown meat and vertical farming pose a promising future to the global food system. Their potential success will largely depend on the creation and implementation of policies that support sustainable and equitable food systems. The successful adoption of these innovations into the food system requires effective collaboration between policymakers, private sectors, and the public to promote policies that improve access to healthy food, reduce environmental impacts, support economic development, and promote technological innovation. Policymakers have an essential role to play in ensuring that these developments do not exacerbate existing inequities in our food system and help to build a resilient and equitable food supply for generations to come.

THE POTENTIAL CHALLENGES IN IMPLEMENTING EFFECTIVE POLICIES

Despite the promises of lab-grown meat and vertical farming, there are several potential challenges that could hinder their full implementation. One of the biggest challenges is the cost associated with these methods of food production. Lab-grown meat, for instance, requires significant investment in research and development, as well as technology, and this will inevitably be reflected in the cost of the final product. This could mean that lab-grown meat, at least initially, will be beyond the reach of many consumers or may be restricted to certain niches of the market. This could slow down the adoption of lab-grown meat and limit its overall impact on food sustainability and security.

Similarly, vertical farming also comes with its own challenges. The high capital costs required to set up vertical farms, as well as the ongoing operational costs, can make it difficult to justify their economic viability. This could then lead to an inability to scale up this farming method and make it accessible to a larger population. There is a risk that vertical farming may not be suitable for certain crops, which could limit its applicability and overall effectiveness. Both lab-grown meat and vertical farming could face regulatory hurdles. In the case of lab-grown meat, there are significant regulatory questions around the safety and ethical considerations of producing meat in this way. Regulators will need to determine whether it is safe to consume meat that has been produced in a laboratory, and whether it meets the same quality standards as traditionally farmed meat. There will also be ethical considerations to consider such as the treatment of animals used in the research for lab-grown meat. All of these factors will require significant research, and if regulations are not well defined, this could limit the future adoption of lab-grown meat. For vertical farming, there are also regulatory issues that need to be addressed. In many jurisdictions, vertical farming is seen as a new technology with new risks that require additional regulations. This may include regulations around the construction and use of the buildings where the crops are grown, as well as regulations around the use of fertilizers and pesticides. As such, there could be challenges in navigating the regulatory landscape and obtaining the necessary licenses and permits to operate a vertical farm. Another potential challenge is the social and cultural acceptance of these innovations. Lab-grown meat, for instance, challenges our traditional understanding of what constitutes meat and could potentially face resistance from those who question the authenticity or safety of consuming laboratory-made meat. Similarly, vertical farming is a significant departure from the traditional farming methods that have been part of our cultural history. This could create social and cultural barriers to the acceptance of this method of farming, particularly in rural communities where farming is a way of life. In addition to these challenges, there are also potential environmental concerns that need to be taken into account. While lab-grown meat and vertical farming offer the potential for significant reductions in greenhouse gas emissions and land use, there are other environmental considerations that need to be taken into account. For example, the energy requirements of these methods may be quite high, which could make them less sustainable in areas where renewable energy is not readily available. Similarly, the disposal of waste products from lab-grown meat production could create environmental issues that need to be addressed.

There are potential economic impacts that could also constrain the adoption of these innovations. For example, lab-grown meat could disrupt traditional livestock farming, potentially leading to job losses and impacts on rural communities where livestock farming is a critical source of employment. Similarly, vertical farming could impact traditional agriculture, leading to changes in the supply chain and potential job losses in this sector. While these impacts may be offset by the potential economic benefits of these innovations, they nonetheless need to be taken into account. While lab-grown meat and vertical farming offer significant promise in terms of increasing food sustainability and security, there are several potential challenges that need to be taken into account. These include cost, regulatory hurdles, social and cultural barriers, environmental considerations, and economic impacts. All of these factors need to be carefully considered in order to ensure that these innovations can be fully realized and leveraged to address the future of food production. By addressing these challenges thoughtfully and proactively, it is possible to pave the way for a more sustainable and secure future for our food supply. The world of food production is transforming as we know it, and two of the most innovative and promising ideas that have emerged in recent years are lab-grown meat and vertical farming. These two concepts are transforming the way we think about food and the future of agriculture. They are part of a revolution that is starting to take place in the world of food production and that has the potential to change the way we eat in important ways. Lab-grown meat has the potential to address problems such as sustainability and animal welfare, while vertical farming has the potential to dramatically reduce the amount of land needed for agriculture and mitigate issues related to climate change, water scarcity, and food security. Lab-grown meat is a revolutionary development in the food industry, and it has the potential to address some of the most pressing problems facing the world today. In traditional meat production, animals are raised in factory farms, which are often associated with animal welfare issues and environmental concerns. Lab-grown meat, on the other hand, is created in a laboratory from animal cells, without the need to raise and slaughter whole animals. This approach provides an opportunity to greatly reduce the environmental impact of meat production, as well as the ethical concerns surrounding animal welfare. By creating meat from animal cells, we can drastically reduce greenhouse gas emissions, land use, and water use associated with meat production. Lab-grown meat has the potential to reduce the spread of animal-borne diseases, such as bird flu or swine flu, by eliminating the need to raise whole animals in close proximity to each other. While lab-grown meat is still in its infancy, researchers and entrepreneurs are rapidly working to perfect the technology, and some companies have even started to bring products to market. Should lab-grown meat become a viable and affordable option, it could have a dramatic impact on the way we eat, providing us with an ethical, sustainable, and healthy source of protein. Vertical farming is another innovative development in the world of food production that has the potential to transform the way we eat. Vertical farming is a method of growing crops indoors, often in multistory buildings, using hydroponic or aeroponic technology. This approach to farming allows us to produce fresh vegetables and fruits year-round, regardless of weather conditions, and reduces the carbon emissions associated with transporting food long distances. Vertical farming also allows us to grow crops in urban areas, reducing the distance between farm and consumer and providing fresh, healthy food to communities that might otherwise lack access to it. Vertical farming can significantly reduce the amount of water and land needed for agriculture, making it a promising solution to issues related to climate change and food security. By growing crops in a controlled environment, we can eliminate the need for harmful pesticides and herbicides and ensure that crops are not contaminated by pollutants or pathogens. While the cost of setting up a vertical farm can be high, the potential for a reliable, year-round supply of fresh produce and the savings in transportation costs and environmental impact make it a promising option for the future of food production. Both lab-grown meat and vertical farming have the potential to revolutionize the way we eat. They offer solutions to many of the problems associated with traditional agriculture, such as land use, water use, and environmental impact. They also offer the potential for increased food security and access to fresh produce, particularly in urban areas. They are not without their challenges. The technology required for lab-grown meat is still in its early stages, and it may be some time before this becomes a viable and affordable option for consumers. The regulatory landscape for lab-grown meat is also uncertain, and it is likely to be the subject of debate and controversy in years to come. Vertical farming requires significant investment in infrastructure and technology, and the energy requirements for these facilities can be high, potentially counteracting some of the environmental benefits.

It is worth considering the potential ethical issues associated with lab-grown meat and vertical farming. While these methods have the potential to address some of the ethical concerns related to traditional agriculture, they may also raise new ethical questions. For example, should we be interfering with the natural order of things by creating meat in a laboratory? And what about the potential for unintended consequences, such as the emergence of new diseases or environmental disruption caused by large-scale vertical farms? It is important to consider these issues as we move forward with these promising innovations.

The future of food is changing, and lab-grown meat and vertical farming are two of the most exciting and innovative developments in the field. These concepts have the potential to dramatically reduce the environmental impact of food production, increase food security, and improve animal welfare. While there are still many challenges to be overcome, and ethical questions to be addressed, these ideas have the potential to transform the way we eat and to create a more sustainable and equitable food system for generations to come.

XII. THE ROLE OF EDUCATION IN THE FUTURE OF FOOD PRODUCTION

As the world population grows at an unprecedented rate, the demand for food is increasing exponentially. To feed the estimated 9.7 billion people by 2050, we need to produce more food in the next 30 years than we have in all of human history. Addressing these challenges in food supply and access requires not only technological advancements but also education, particularly in the area of sustainable agriculture. Education plays a vital role in providing knowledge and skills necessary to develop and implement sustainable practices in food production. With new technologies emerging, teaching key principles of resource and energy conservation, recycling, and waste management can be a game-changer. Through education and innovation, we can work toward a more sustainable future in agriculture.

One of the technologies that could drive this change is vertical farming. By growing crops in vertically stacked layers, plants can be grown in a fraction of the space required for conventional agriculture, using 70-95% less water than traditional farming. Vertical farming also allows for year-round crop production, with much greater control over climate, lighting, and nutrient supply. By incorporating intelligent systems and sensors, vertical farming can reduce energy costs and improve crop yield and quality. To ensure the success of vertical farming, however, it is crucial to educate farmers and the public about the technology and its associated benefits and challenges. Agricultural education institutions can equip students with the knowledge and skills to design and operate vertical farms, providing a sustainable solution to food production in urban and rural areas alike. By educating the next generation of farmers in this technology, we can help build a more sustainable and resilient food system, which is essential for meeting future food demands. Educational institutions should invest in research to develop advanced technologies that can help promote the adoption of vertical farming practices. This includes sensors for monitoring crop growth and nutrient requirements, as well as innovative solutions for lighting and energy use. Another technology that could revolutionize the food industry is lab-grown meat. This innovation addresses sustainability and ethical concerns surrounding traditional animal agriculture. Lab-grown meat involves growing animal muscle cells in a laboratory to produce meat without the need for livestock breeding and slaughter. It offers a solution for producing protein without using vast amounts of natural resources, such as water, space, and animal feed. There is a growing demand for more sustainable and ethical food production. With the adverse effects of animal agriculture on the environment becoming increasingly apparent, consumers are looking for alternatives. Education can play a significant role in promoting the use of lab-grown meat as a solution. Educational institutions can help to familiarize the public with the technology, dispelling any myths and addressing any potential concerns. They can invest in research, helping to establish protocols for the sustainable production of lab-grown meat. Another way education can promote sustainability in food production is by emphasizing the importance of reducing food waste. The United Nations estimates that one-third of the food produced globally is lost or wasted each year. This represents significant inefficiencies in our food system, with the economic, social, and environmental impacts of food waste being substantial. Education can help to reduce food waste by promoting awareness of the problem and encouraging responsible consumption. One way to achieve this is through educating individuals on the storage and handling of food, helping to extend the lifespan of products. Educational institutions can teach the principles of circular business models, such as the conversion of food waste into biogas, compost, or animal feed. By teaching sustainable practices in food production and consumption, we can cultivate a more informed and responsible approach to food waste reduction. Education plays a critical role in the future of food production. With the challenges facing our food system becoming increasingly apparent, it is essential to equip the next generation of farmers and consumers with the knowledge and skills to develop sustainable practices. This includes promoting awareness and adoption of innovative technologies such as vertical farming and lab-grown meat, both of which offer sustainable solutions to the challenges we face in the food industry. Emphasizing the principles of circular business models and reducing food waste is also essential. By incorporating sustainability into the curriculum of educational institutions, we can work towards a more resilient, equitable, and sustainable food system for generations to come.

THE NEED FOR EDUCATION ON LAB-GROWN MEAT AND VERTICAL FARMING

The rise of lab-grown meat and vertical farming has sparked the need for education on these new technologies. While they present great potential in terms of solving the problems of food shortages and environmental degradation, they also represent completely new ways of sourcing food. Lab-grown meat, for example, is created by cultivating muscle cells in a laboratory setting, while vertical farming involves growing crops in stacked layers, often inside urban buildings. The novelty of these methods calls for a deeper understanding of their implications, and an introduction to these topics should be included in educational curriculums at all levels. Education on lab-grown meat and vertical farming can also contribute to further innovation and development in these fields. One of the most pressing reasons to educate people on these innovations is their potential to address global food insecurity, which is predicted to worsen in the coming decades. According to the United Nations, the world's population is projected to reach 9.7 billion people by 2050 and will require a 70% increase in global food production to meet the needs of everyone. The resources needed to produce traditional meat, such as water and land, are becoming increasingly scarce, and climate change is further exacerbating these issues. Lab-grown meat and vertical farming, on the other hand, have the capacity to produce food in a more efficient and sustainable manner. In the case of lab-grown meat, it requires up to 99% less land, 96% less water, and produces 96% less greenhouse gas emissions than conventional livestock farming. Similarly, vertical farming uses up to 90% less water and land than traditional farming and can reduce transportation costs due to their placement in urban areas. Educating people on the potential benefits of these innovations, particularly in terms of global food security, can encourage their adoption and further development.

Another crucial aspect of educating people on lab-grown meat and vertical farming is to address the ethical concerns that surround conventional animal agriculture. With lab-grown meat, animals are not slaughtered for food, and it eliminates animal suffering and reduces animal rights debates associated with conventional agriculture. Some ethical concerns still exist around the use of fetal bovine serum (FBS), a growth medium that originates from the blood of cow fetuses, in the cultivation process. Educating people on the ethics of lab-grown meat, and the methods that are being explored to create alternatives to FBS, could create important discussions around animal rights and lead to greater acceptance and use of lab-grown meat.

Similarly, vertical farming can also address ethical concerns around animal agriculture, as it is a plant-based product. By providing fresh, local, and organic food, the practice of vertical farming can also support the health and welfare of humans and the planet as a whole. Educating people on the ethics of vertical farming and its practices can create greater awareness and foster its adoption. In order for these innovations to be successful, it is essential to have a diverse and educated workforce that can support their development and implementation. Education on lab-grown meat and vertical farming can create new job opportunities and provide training for future generations to enter these fields. Hands-on training in labs and vertical farms would provide students with the practical experience necessary for a career in these industries. Educational programs could include courses on the policy and regulatory frameworks that govern these industries, such as intellectual property law, food safety regulations, and environmental protection laws.

Education on lab-grown meat and vertical farming can also promote further innovation and development in these fields. By increasing awareness and understanding of these concepts, more problems can be identified and solved. Researchers and scientists can explore new technologies and improve existing ones. This can lead to the development of new innovations that can transform our agricultural systems even further. With more knowledge and competition, there could be more investment in these industries, thereby leading to greater innovation, efficiency, and sustainability in the future. As the world faces increasing challenges around food security and environmental degradation, the need to educate people on the innovations of laboratory-grown meat and vertical farming has never been more critical. We need to educate individuals at all levels of society so that they can make informed decisions about the role of these innovations in our world. Educational efforts will allow us to increase awareness, spur development, and lead to successful adoption of these technologies. It is essential that policymakers, educators, and industry leaders collaborate to promote innovation, create jobs, and ensure that these technologies will be inclusive and accessible for all.

THE POTENTIAL ROLE OF EDUCATION IN PROMOTING SUSTAINABLE AND EQUITABLE FOOD SYSTEMS

Education can and should play a crucial role in promoting sustainable and equitable food systems. By educating individuals and communities on the importance of sustainable food production and consumption, we can shift attitudes towards food, create demand for sustainable and local food options, and ultimately, reduce the environmental impact of food production. Education can also raise awareness about the social and economic factors that contribute to food insecurity and inequality and empower individuals and communities to take action to address these issues. One key aspect of educating individuals on sustainable food systems is promoting understanding of the environmental impact of food production. The global food system accounts for a significant portion of greenhouse gas emissions, water consumption, and land use, which all contribute to climate change and environmental degradation. By providing individuals with knowledge and tools to incorporate sustainable practices into their daily lives, we can help reduce these impacts. For example, individuals can learn about the benefits of reducing meat consumption, choosing plant-based options, and supporting locally grown and sourced food. Educational campaigns can also focus on reducing food waste, creating composting systems and encouraging low-impact transportation options for food. Through initiatives such as farm-to-school programs, urban agriculture, and community gardens, we can also educate individuals on the importance of local food production and its benefits for the environment and the local economy. In addition to environmental sustainability, education can also promote equity in the food system. Despite producing enough food to feed the global population, there are still millions of people who suffer from hunger and malnutrition. Factors such as poverty, lack of access to education and resources, and unequal distribution of food contribute to these issues. Education can help individuals and communities identify and address these systemic factors, and work towards creating more equitable and resilient food systems. Programs that promote food sovereignty, such as community-led agriculture, can empower communities to take control of their food production and distribution systems. Initiatives that promote education on nutrition and healthy eating can help improve health outcomes and reduce health disparities in marginalized communities. Education can also play a crucial role in shaping policy and governance around food systems. By educating individuals and communities on the importance of sustainable and equitable food systems, we can create demand for policy changes that align with these goals. Through advocacy, community organizing, lobbying and education, groups can push for policies that align with their vision of a more sustainable, equitable, and resilient food system. This can include policies that promote sustainable agriculture, support small-scale and family-run farms, regulate the use of pesticides and fertilizers, and incentivize the production and distribution of local and organic food. Education can also encourage innovation in the food system. As technologies such as lab-grown meat and vertical farming continue to emerge, educating the public on the potential benefits and drawbacks of these innovations can help ensure that they are incorporated into the food system in a sustainable and responsible way. Education can also promote creativity and collaboration, encouraging individuals and communities to develop innovative solutions to food system challenges. For example, community-led agriculture projects, such as community supported agriculture (CSA), can provide a platform for experimentation and innovation in sustainable agriculture practices.

Education can play a crucial role in promoting sustainable and equitable food systems. By providing individuals and communities with knowledge and tools to incorporate sustainable practices into their daily lives, we can reduce the environmental impact of food production and create demand or sustainable and local food options. Education can also raise awareness about the social and economic factors that contribute to food insecurity and inequality and empower individuals and communities to take action to address these issues. By shaping policy and governance, encouraging innovation, and promoting equity in the food system, education can help create a more sustainable, equitable, and resilient food system for present and future generations.

THE POTENTIAL CHALLENGES IN IMPLEMENTING AN EFFECTIVE EDUCATION FRAMEWORK

The primary challenge in implementing an effective education framework is the need to balance the needs of all stakeholders. From government officials to educators, parents, and students, each group has their own priorities, preferences, and expectations for the education system. For instance, government officials might prioritize funding classroom technology such as tablets and laptops, while educators may believe that traditional teaching methods such as chalk and blackboards are more effective. Parents may value extracurricular activities such as sports and music, while students might prefer more flexible schedules or vocational training. These competing interests may diffuse focus away from establishing a comprehensive education system that caters to the needs of students. Another challenge involves the cost of retraining teachers and providing new resources. As education systems evolve, teachers must adapt to new pedagogies and innovations. Although some institutions might be inclined to invest in their teachers' professional development, others may not have the necessary resources or political will to support upskilling efforts. With advancements in technology, there may be a need to continuously invest in modern resources and tools, such as new teaching software, virtual reality classrooms, or augmented reality educational tools, that may not fit traditional budgets. There is a need to address inequities to ensure that education opportunities are available to all students. A significant proportion of students may not have equal or equitable access to education resources, opportunities, or facilities. For instance, students in economically deprived areas may not have access to the same resources, such as well-staffed libraries, textbooks, laboratories, support services, or afterschool activities, as their peers in more affluent areas. The digital divide also presents a roadblock, with students from low-income families less equipped to leverage online resources. This lack of resources can negatively impact academic performance and perpetuate social and economic disparities. Cultural and linguistic barriers may pose additional difficulties for students who are not proficient in the language of instruction or are part of marginalized communities. Educators must focus on identifying and addressing these challenges to ensure that every student has an equal opportunity for success. There are issues related to the lack of data-driven education policies which can result in ineffective reform. Despite the phenomenon of big data, lack of an evidence-base underlies many education policies and strategies. Without a reliable understanding of the extent and nature of various learning challenges, policy-makers may take unnecessary risks while implementing changes in the education systems. For instance, an overemphasis on standardized test scores can lead to undue pressure on students and a narrow focus on testing that ultimately erodes quality education. In an era of alternative facts and unreliable news, innovative education reforms require a systematic approach to evidence-based policymaking. Educators and policy-makers must rely on concrete evidence in developing education policies, interventions, and programs. Thus, it is important to build a consensus on best practices through collaborations with educators and researchers in the field, input from policymakers, and collaboration with education technology companies to build effective tools that can be evaluated through empirical research. Implementing effective education frameworks is a complex and multi-faceted endeavor. It involves providing high-quality, accessible, and innovative education to all students while balancing the competing interests of multiple stakeholders. Nonetheless, with the persistent efforts of educators, researchers, policymakers, and the broader community, it is possible to develop and maintain high standards of education that can prepare students to meet the challenges of the future. To this end, it is critical to continuously invest in professional development for educators, utilize evidence-based policy-making, and promote equitable access to education resources across all segments of society. Through such concerted efforts, the education sector can overcome potential challenges and provide the kind of high-quality education that is needed to pave the way for a brighter tomorrow. Two of the most promising innovations in food production today are lab-grown meat and vertical farming. These technologies may represent the future of food as they offer a sustainable and more efficient way to produce protein-rich foods. Besides, they could help mitigate some of the adverse environmental impacts associated with traditional meat production and agricultural practices. Lab-grown meat, also known as cell-based meat, is produced by cultivating muscle cells in the lab and providing them with nutrients and growth factors. This technique resembles the way traditional meat is produced biologically, but without the need for animal rearing, slaughter, or the use of antibiotics or growth hormones. The result is a product that is genetically identical to meat but does not require the same resources or entail as many ethical or environmental concerns. Vertical farming is a method of growing crops in stacked layers or shelves using artificial lighting, temperature, and humidity controls. This technique enables year-round crop production, closer proximity to urban centers and consumers, and significatively reduces water consumption, land use, and greenhouse gas emissions. Together, these technologies offer a new vision of the food system that could be more sustainable, secure, and equitable. The main advantage of lab-grown meat is that it could potentially reduce the environmental impact of conventional livestock production. Traditional meat production relies on vast amounts of resources such as water, land, and feed, and causes serious environmental problems such as deforestation, water pollution, and greenhouse gas emissions. For example, one kilogram of beef requires twenty times more land and water than the same amount of soy or grains. Animal agriculture is responsible for fourteen percent of global greenhouse gas emissions, which is more than the entire transport sector combined. Replacing animal-based meat with lab-grown meat could help to address some of these issues. According to a study by the University of Oxford, lab-grown meat could reduce greenhouse gas emissions by up to ninety-five percent, use ninety-nine percent less land, and up to ninety-six percent less water than traditional meat. As a result, lab-grown meat could be a critical tool for mitigating the worst effects of climate change, ensuring food security, and meeting the growing demand for food in an era of population growth. Another advantage of lab-grown meat is that it could lessen some animal welfare concerns related to traditional meat production. Traditional livestock production involves practices such as the use of antibiotics and growth hormones, confinement, and slaughter methods that cause stress and pain to animals. The production of lab-grown meat eliminates these concerns since it does not require animal rearing and slaughter. It may also be possible to produce meat products that are free from diseases and contaminants such as Salmonella or E. Coli. By providing an alternative source of meat, lab-grown meat could also help to reduce the demand for meat from factory farms where animal welfare standards are often poor and raise public consciousness about animal rights. Vertical farming can also revolutionize the food system by making it more efficient, sustainable, and resilient. This method allows for year-round crop production by optimizing the use of space, water, and nutrients. It also reduces the distance between production and consumption centers, which cuts transportation costs, increases freshness, and reduces food waste. Vertical farming systems also require less water and use fewer pesticides and fertilizers, which can significantly reduce the vast amounts of water and chemicals used in traditional agriculture. Vertical farms can be located in abandoned warehouses, urban rooftops, or unused land, bringing fresh products closer to consumers and reducing the need for long-distance transportation and refrigeration. Vertical farming also offers opportunities to experiment with new cultivars of crops, provide local and fresh products to consumers, and grow non-traditional crops such as medicinal plants or edible insects. As an adaptable farming technique, vertical farming offers many benefits, including a reduction in resource use, a more consistent harvest, and fewer pests and diseases. By facilitating new and sustainable ways to grow fresh produce, vertical farming can be a powerful tool for improving the food security of urban populations or communities living in areas with limited access to fresh and nutritious foods. Lab-grown meat and vertical farming represent two promising innovations that may transform the future of food. There still are some challenges to overcome, such as scaling up production, reducing costs, and overcoming regulatory and consumer uncertainties. Nonetheless, these technologies offer significant benefits for the environment, animal welfare, and food security. As global food demand continues to rise, and the planet undergoes climate change, innovations in food production such as lab-grown meat and vertical farming offer an exciting glimpse of a more sustainable and equitable future.

XIII. THE ROLE OF CONSUMERS IN THE FUTURE OF FOOD PRODUCTION

The innovations in food production, such as lab-grown meat and vertical farming, have the potential to change the way we eat. The success of these innovations largely depends on consumer acceptance and demand. The role of consumers in the future of food production is crucial in shaping the market and influencing the sustainability and ethics of food production. Consumers have the power to choose what they eat and influence the food industry by demanding transparency and accountability in food labeling and production practices. Firstly, consumers can drive the demand for sustainable and ethical food production practices. The growing concern over the negative environmental impact of conventional animal agriculture has led to increased interest in alternative protein sources. Lab-grown meat has emerged as a promising technology that can provide an alternative to conventional animal agriculture, with lower greenhouse gas emissions and less land and water use. If consumers are willing to try lab-grown meat and shift their consumption habits towards more sustainable and ethical food sources, this could incentivize the food industry to invest more in these technologies and ultimately shift towards a more sustainable and ethical food system. Similarly, consumers can also influence the adoption of vertical farming as a more sustainable and efficient way to produce crops. Vertical farming uses stacked layers of crops grown under controlled conditions, such as hydroponics or aeroponics, without the need for pesticides or herbicides. As vertical farms can be located in urban areas, they can also reduce the carbon footprint associated with transportation and refrigeration of crops. The success of vertical farming depends on consumer demand for locally sourced and sustainably grown produce. If consumers are willing to pay a premium for fresh and sustainable produce and demand more transparency in food labeling, this could incentivize the growth of vertical farms and ultimately improve the sustainability and resilience of our food system.

The adoption of lab-grown meat and vertical farming faces challenges with the perception and acceptance of these technologies among consumers. For instance, lab-grown meat still faces the challenge of being perceived as unnatural or less tasty than conventional meat. Lab-grown meat is currently more expensive to produce than conventional meat, which could deter consumers from adopting it as a regular food source. Similarly, the adoption of vertical farming depends on overcoming the perception that it is a niche or expensive technology and that conventional farming is still a better option. Educating and engaging consumers on the benefits and potential of these technologies is essential in driving consumer demand, increasing acceptance and ultimately advancing the future of food production.

Consumers can also play a role in shaping the ethics and transparency of food production by demanding clear labeling and information on the production practices of food products. The increasing concern over animal welfare and the use of antibiotics and hormones in conventional animal agriculture has led to consumers demanding more transparency and ethical practices in animal products. For instance, the labeling of free-range or cage-free poultry and eggs has become a popular way for consumers to identify and choose more ethical animal products.

Consumers are increasingly interested in knowing the origin and production practices of their food, with certifications like Fairtrade and Organic gaining popularity. This demand for transparency and ethical practices can incentivize the food industry to adopt more sustainable and ethical production practices and improve the welfare of animals and the environment.

The role of consumers in the future of food production also includes the responsibility of reducing food waste, promoting food security, and supporting local and sustainable food systems. The current global food system is responsible for significant greenhouse gas emissions and contributes to biodiversity loss, water scarcity, and soil degradation. Reducing food waste through responsible consumption habits and supporting local and sustainable food systems can help mitigate these negative impacts. Supporting small-scale farmers and local food systems can promote food security, reduce the carbon footprint associated with transportation and distribution and support local communities.

The innovations in food production, such as lab-grown meat and vertical farming, have the potential to transform the way we eat and address some of the pressing sustainability challenges of our current food system. The success of these technologies depends on the acceptance and demand of consumers. Consumers play a crucial role in shaping the future of food production by promoting sustainable and ethical food sources, demanding transparency and accountability in food labeling and production practices, reducing food waste, and supporting local and sustainable food systems. The future of food production is not only a technological or scientific challenge, but also a social and cultural challenge that requires the active participation and engagement of consumers.

THE POTENTIAL IMPACT OF CONSUMER BEHAVIOR ON THE SUCCESS OF LAB-GROWN MEAT AND VERTICAL FARMING

As exciting as the innovations in lab-grown meat and vertical farming may be, their success ultimately comes down to consumer adoption. In the case of lab-grown meat, one critical factor is whether consumers will be willing to eat it in the first place. Some may balk at the notion of meat that isn't "real," or may have concerns about the safety of such products. Gaining consumer trust will be vital for the success of lab-grown meat, and this will necessitate extensive education and marketing efforts. One thing working in favor of lab-grown meat is its potential to be more sustainable than traditional animal agriculture. By eliminating the need for large-scale livestock farming, lab-grown meat could reduce land use, water consumption, and greenhouse gas emissions. If consumers are motivated by environmental concerns, this could be a major selling point. The ability to custom-tailor the nutritional content of lab-grown meat could also be a draw for health-conscious consumers. There are challenges to producing lab-grown meat on a large scale. Currently, the process is time-consuming and expensive, and there are concerns about the scalability of production. The success of lab-grown meat will depend not only on consumer demand but also on continued innovation and investment in the technology.

As for vertical farming, the potential benefits in terms of sustainability and food security are clear. By growing crops in controlled indoor environments, vertical farms can reduce water use, minimize pesticide use, and increase food production in urban areas. The ability to grow crops year-round, regardless of weather conditions, is another major advantage. As with lab-grown meat, the success of vertical farming will depend heavily on consumer acceptance. There may be some skepticism about the nutritional quality and taste of fruits and vegetables grown in indoor environments, rather than in soil. Vertical farming produce may be more expensive than conventionally grown produce, which could limit its appeal to some consumers. To address these challenges, marketing efforts will be key. Proponents of vertical farming will need to educate consumers about the potential benefits of the technology, emphasizing the sustainability and security implications. They may also need to experiment with different pricing strategies, such as offering lower prices for bulk purchases or partnering with grocery stores to offer exclusive deals. Another factor that could impact the success of vertical farming is the regulatory environment. There may be concerns about food safety and quality control, particularly if vertical farms are seen as untested or unregulated. While there are currently no federal regulations in the United States specifically addressing vertical farming, this could change if the technology gains more widespread adoption. The potential impact of consumer behavior on the success of lab-grown meat and vertical farming is significant. While both technologies offer exciting possibilities for the future of food production, their long-term viability will depend on widespread consumer adoption. In the case of lab-grown meat, this will require building consumer trust through education and marketing efforts, as well as addressing concerns about safety and scalability. Meanwhile, in the case of vertical farming, proponents will need to convince consumers of the nutritional quality and taste of their products, while also addressing pricing and regulatory concerns.

If these challenges can be overcome, the potential benefits of lab-grown meat and vertical farming are immense. By offering more sustainable and secure food production methods, these technologies could play a critical role in addressing some of the most pressing challenges facing our food system. The adoption of these innovations could also lead to more localized and community-based food systems, as urban areas become more self-sufficient in their food production. As we look to the future of food, it is clear that lab-grown meat and vertical farming will play an increasingly important role, and that the decisions and behaviors of consumers will be a critical factor in determining their success.

THE POTENTIAL BENEFITS OF EDUCATED CONSUMER CHOICES

In the midst of all these technological innovations in food production, there is also a growing interest in "educated consumer choices" - that is, consumers who are informed about the risks and benefits of various food options, and who make more conscious decisions about what to eat and how to source their food. There are a number of potential benefits to this kind of approach, both for consumers themselves and for the broader food system as a whole. One of the main benefits of educated consumer choices is that they can have a positive impact on public health. For example, by understanding the nutritional value of different foods, consumers can make more informed decisions about what to eat in order to maintain a balanced diet. As consumers become more aware of the negative health impacts associated with certain types of food (such as foods high in sugar or saturated fat), they may choose to limit their consumption of these products, which could in turn reduce rates of obesity, diabetes, and other diet-related illnesses. Another possible benefit of educated consumer choices is that they can help to promote more sustainable food practices. For example, by choosing to buy food from local farmers or co-ops rather than large multinational corporations, consumers can support smaller-scale, more environmentally-friendly farming operations. Similarly, by avoiding foods that are produced using environmentally harmful practices (such as factory farming or monoculture), consumers can help to reduce their own carbon footprint and contribute to a more sustainable food system overall. In addition to these health and sustainability benefits, educated consumer choices can also help to promote social justice and equity within the food system. By choosing to support small-scale farmers and businesses, consumers can help to ensure that more of their food dollars go towards supporting local communities rather than just enriching large corporations. By becoming more informed about food labeling and marketing practices, consumers can be better equipped to identify deceptive or misleading claims about food products, which can help to prevent consumers from being exploited by unscrupulous marketers. So how can consumers become more educated and empowered in their food choices? One approach is to advocate for increased transparency and labeling requirements within the food industry. For example, many consumers are now demanding more detailed nutritional information, such as the total amount of sugar or sodium in a given product, be included on food labels. There is growing interest in more expansive labeling requirements that would require companies to disclose the environmental impact of their production processes and other relevant sustainability metrics. Another approach to promoting educated consumer choices is through food education programs and initiatives. Schools, community groups, and other organizations can help to teach consumers about nutrition, sustainability, and other related topics, as well as providing resources and support for consumers who are interested in making more conscious food choices. For example, some organizations have started community gardens and urban farming initiatives, allowing local residents to learn about sustainable farming practices and grow their own food in the process.

The potential benefits of educated consumer choices are

numerous and far-reaching. By making more informed decisions about what we eat and how we source our food, we can help to promote better public health, a more sustainable food system, and greater social and economic equity within our communities. As the food industry continues to evolve and new innovations emerge, it is more important than ever for consumers to be actively engaged and informed about the food choices they make.

THE POTENTIAL CHALLENGES IN CHANGING CONSUMER BEHAVIOR PATTERNS

While the idea of lab-grown meat and vertical farming has the potential to revolutionize food production, it is important to consider the challenges that these innovations may face in changing consumer behavior patterns. One of the biggest challenges in this regard is the fact that many consumers may be resistant to trying new and unfamiliar products. Even if these products are more sustainable and environmentally friendly than traditional meat sources, some consumers may still be hesitant to try them due to concerns about taste, texture, and potential health effects. This reluctance to try new foods may be even more pronounced in cultures where meat is a particularly important part of the diet, such as in the United States, where meat consumption is high and often associated with cultural identity.

Another potential challenge in changing consumer behavior patterns is the fact that lab-grown meat and vertical farming may be seen as unnatural or "fake" by some consumers. This perception may be particularly strong among consumers who are concerned about the use of biotechnology and genetic engineering in food production. While lab-grown meat and vertical farming do not necessarily involve genetic engineering, they are still relatively new and unfamiliar technologies that may be met with skepticism or even fear by some consumers. In order to overcome these concerns, it will be important to focus on educating consumers about the benefits and safety of these technologies, as well as the potential environmental and animal welfare benefits that they offer. Another potential challenge in changing consumer behavior patterns is the fact that many consumers may not be willing or able to pay a premium price for lab-grown meat and vegetables grown in vertical farms. While proponents of these technologies argue that they will ultimately be cost-effective in the long run, there may be an initial cost barrier to overcome in terms of both production and consumer pricing. This could be particularly challenging in developing countries, where food security and access to affordable nutrition are major concerns. High-end consumers, who might be most interested in lab-grown meat, may also be concerned about the potential social stigma of consuming an alternative meat product, which might be seen as a substitute for meat, rather than a choice of preference. In addition to these potential challenges, there are also some practical logistical issues that will need to be addressed in order to successfully implement lab-grown meat and vertical farming. For example, vertical farming requires a significant amount of space and resources, such as water and energy. This may be a particular challenge in urban areas, where space is at a premium and resources are often scarce. Scaling these technologies up to meet the needs of a growing global population may also prove difficult, as production and distribution may be costly and difficult to manage at a large scale. Another logistical challenge may be the need for new regulations and standards to govern the safety and quality of these products, particularly in the absence of existing regulatory frameworks that may not be applicable to these new types of food products. In addition to these logistical and practical challenges, there may be broader social and cultural factors that could make it difficult to change consumer behavior patterns. Some of these factors may include deeply ingrained cultural beliefs about the importance of traditional diets and the role of meat in these diets, as well as religious and ethical concerns related to the consumption of meat. In order to overcome these barriers, it will be important to engage with communities and consumers in a constructive and positive way, emphasizing the many benefits of these technologies for both human health and environmental sustainability. While the potential benefits of lab-grown meat and vertical farming are clear, there are also many challenges that will need to be overcome in order to successfully change consumer behavior patterns. These challenges may include resistance to new and unfamiliar products, concerns about the safety and naturalness of these technologies, cost barriers, logistical and regulatory issues, and broader social and cultural factors. With time and effort, it is possible to overcome these challenges and create a more sustainable, healthy, and equitable food system for all. Lab-grown meat and vertical farming are two major innovations in the field of food production that have the potential to revolutionize the way we eat and sustain ourselves in the future. These groundbreaking developments are driven by the pressing need to address the challenges and limitations of traditional farming practices and meat production systems that are unsustainable, resource-intensive, and often unethical. Both lab-grown meat and vertical farming offer radical alternatives that promise to reduce environmental impact, conserve resources, increase food security, and improve public health. Despite the current skepticism and controversies surrounding these new methods, many experts believe that they hold great promise for the future of food production and consumption. Lab-grown meat, also known as cultured or cell-based meat, involves using animal cells to produce meat products in a laboratory without the need for animal slaughter or rearing. The process involves taking a small sample of animal tissue and culturing it in a controlled environment with nutrients and growth factors that stimulate the cells to multiply and form muscle tissues. The result is a flesh-like substance that has the same nutritional composition, flavor, and texture as conventional meat. This method has the potential to address several ethical and environmental issues associated with traditional meat production. For instance, it would eliminate the suffering and exploitation of animals in factory farms, reduce greenhouse gas emissions, and save land and water resources. Cultured meat could also lead to the development of new and healthier meat products that are free from antibiotics, hormones, and other harmful substances that are currently prevalent in conventional meat. There are several challenges to overcome before cultured meat can become a mainstream food item. One of the main hurdles is the high cost of production and the need for large-scale commercialization to bring down prices. Regulatory and social acceptance issues must be addressed to ensure that consumers trust the safety and quality of lab-grown meat. Despite these challenges, several startups and research organizations are investing in this technology and making significant progress in improving the efficiency and scalability of the process.

Vertical farming is another innovation in food production that has gained traction in recent years. It involves growing crops in vertically stacked layers using hydroponic or aeroponic systems that allow for precise control of nutrient and water supply. This method allows for year-round cultivation of crops in controlled environments that are not affected by weather conditions, pests, or soil quality. Vertical farms can also be located in urban areas, reducing the need for long-distance transportation and minimizing the carbon footprint of food production. By using sensors, data analytics, and artificial intelligence, vertical farming can optimize crop yields, reduce resource use, and enhance food safety. Vertical farming has several advantages over traditional agriculture, but it also faces challenges that must be addressed to realize its full potential. One of the main limitations is the high upfront investment required to build and maintain vertical farms, which can deter small-scale farmers and limit access for low-income communities. There are concerns about the energy consumption, waste generation, and overall sustainability of vertical farming systems. To address these challenges, there is a need for innovative financing models, better integration of renewable energy, and continued research on the environmental impact of vertical farming. Despite these challenges, there is growing interest and investment in vertical farming, especially in densely populated cities, where space and resources are limited. Many startups and established companies are developing new models and technologies that aim to make vertical farming more accessible, affordable, and sustainable. For example, some companies are experimenting with vertical farms integrated into residential and commercial buildings, which can reduce costs and energy consumption by using waste heat and natural light sources.

Lab-grown meat and vertical farming are two innovative methods that have the potential to transform the way we produce and consume food in the future. These methods offer solutions to many of the challenges and limitations of traditional farming practices and meat production systems, from resource scarcity to environmental degradation to animal welfare concerns. While both lab-grown meat and vertical farming still face significant barriers to widespread adoption, there is growing interest and investment in these fields, which suggests that they will play a significant role in the future of food production and consumption. As consumers and policymakers become more aware of the benefits and challenges of these innovations, it is essential to engage in informed debates and discussions about their potential impact on society, the environment, and our health. By doing so, we can ensure that these new methods of food production are sustainable, equitable, and aligned with our values and needs.

XIV. CONCLUSION

The future of food is heading towards a change that will revolutionize our relationship with what we eat. The rise of lab-grown meat and vertical farming represents a shift in our thinking about food, where we are no longer dependent on the traditional methods of farming and animal husbandry that have been used for thousands of years. These innovations in food production are driven by the growing population, climate change, and concerns over animal welfare, as well as an increasing demand for sustainable and efficient food sources. The implications of these technologies are significant, not just for our health and wellbeing, but also for the environment and the economy. Lab-grown meat could potentially change the way we view meat consumption, as it has the potential to be healthier, more ethical, and sustainable. By eliminating the need for animal slaughter, it addresses concerns over animal welfare, while also reducing greenhouse gas emissions and other environmental impacts associated with animal agriculture. Lab-grown meat could also be produced in a controlled environment, free from the antibiotics, hormones, and other chemicals that are commonly used in traditional animal farming. It has the potential to provide a more consistent and predictable protein source, which could be used to meet the growing demand for meat in developing countries. Although lab-grown meat is not yet commercially available, it has been well-received by the food industry and has gained interest from investors and consumers alike. The rise of vertical farming is another advancement that has the potential to change the way we grow and consume food. Vertical farms use indoor farming techniques to grow crops in stacked layers, using artificial lighting, temperature control, and nutrient-rich solutions to mimic natural growing conditions. They offer a space-efficient and sustainable way of growing crops, while also reducing the need for pesticides, herbicides, and fertilizers. Vertical farming has the potential to produce a larger amount of crops per acre compared to traditional farming, making it more efficient and cost-effective. As a result, it could help reduce food waste, improve food security, and provide an alternative to food imports. Although there are still many challenges to overcome, including the high costs of technology and research, the potential benefits of lab-grown meat and vertical farming are far-reaching. These innovations have the potential to transform the way we produce and consume food, making it healthier, more ethical, and sustainable. It is important to acknowledge the potential risks associated with their adoption, as well as the limitations of current technology and research. For example, lab-grown meat may require more energy than traditional animal agriculture, while vertical farming may struggle to produce crops that require pollination by insects. People may still have cultural and psychological barriers to accepting lab-grown meat as a food source. Despite these challenges, the future of food looks bright. Innovations such as lab-grown meat and vertical farming have the potential to address some of the most pressing concerns of our time, including food security, climate change, and animal welfare. They offer a more sustainable and efficient way of producing and consuming food, while also promoting healthier eating habits and reducing our impact on the environment. It is up to us, as consumers, investors, and policymakers, to support these developments, while also being mindful of the potential risks and limitations. By working together to create a more sustainable food system, we can ensure that future generations have access to nutritious and affordable food, as well as a healthy and thriving planet.

A SUMMARY OF THE KEY POINTS DISCUSSED IN THE ESSAY

In summary, this essay has made a compelling case for the future of food, which lies in two primary innovations: lab-grown meat and vertical farming. In discussing lab-grown meat, the essay has highlighted the environmental and ethical issues associated with animal agriculture, which lab-grown meat has the potential to address. Lab-grown meat also has the advantage of reducing the risks associated with foodborne illnesses and improving the nutritional quality of meat. The essay has also discussed the state of research and development in lab-grown meat, outlining the challenges and opportunities that lie ahead for this innovation. In discussing vertical farming, the essay has emphasized the potential for this approach to revolutionize food production, especially in urban areas where space is limited. Vertical farming has the advantage of using fewer resources than traditional farming, such as water, fertilizer, and energy, and it also reduces the need for transportation and storage of fresh produce. The essay has also discussed the challenges associated with vertical farming, including the initial costs, the reliance on technology, and the need for automation. The essay has explored the potential impacts of these innovations on the way we eat, highlighting the health, social, and cultural factors that may come into play. All in all, the future of food is exciting, with lab-grown meat and vertical farming offering promising solutions to some of the most pressing challenges facing our food system today.

A CALL TO ACTION FOR POLICYMAKERS, EDUCATORS AND CONSUMERS TO PROMOTE SUSTAINABLE AND EQUITABLE FOOD SYSTEMS

As we embrace the era of innovation, policymakers, educators, and consumers must join hands in promoting sustainable and equitable food systems. The food production industry is complex, diverse, and ever-evolving, making it challenging to create practical policies that ensure sustainability and equity. It is imperative for policymakers to implement legislation that supports sustainable practices and equity in food production. Educators must also be involved in educating consumers and promoting sustainable practices, starting from early education levels. Consumers must play an active role in food production by supporting sustainable practices and demanding transparency from producers. Sustainable and equitable food systems require a multidimensional approach that incorporates economic, social, and environmental factors. To this end, policymakers, educators, and consumers must come together to promote transformative changes in the food production industry. Sustainable food systems prioritize environmental stewardship over profit-making. They minimize waste and pollution while promoting resource conservation. They maintain the ecosystem's balance by supporting soil health, biodiversity, and climate regulation. Sustainability in food production is crucial, given that the current modern food system is responsible for many environmental and social challenges, including deforestation, greenhouse gas emissions, land degradation, and loss of biodiversity. To promote sustainability, policymakers must enact comprehensive legislation that favors sustainable practices over unsustainable ones. For instance, providing incentives for farmers who adopt sustainable practices such as crop rotation, intercropping, and reduced tillage could mitigate soil degradation and promote soil health. Policymakers should advocate for increased access to information and technical assistance for farmers to help them adopt sustainable practices. Equitable food systems, on the other hand, prioritize fair food access and production. Food equity aims at ensuring that everyone has access to safe, healthy, and culturally appropriate food. It also promotes fair labor practices in the food production industry. An equitable food system considers all the stakeholders involved in food production, including farmers, workers, and consumers. Equitable food systems can mitigate social inequalities stemming from the modern food system, such as food deserts, lack of access to healthy food, and labor exploitation. Policymakers must advocate for legislation that promotes fair labor practices, such as fair and safe working conditions, living wages, and access to benefits for food system workers, farmers, and food industry workers. They must also increase access to healthy and fresh food through community gardens, farmers markets, and other initiatives that bring healthy food to underserved communities. Educators play a critical role in promoting sustainable and equitable food systems. By educating consumers on sustainable food practices, they can foster a culture of sustainability and create informed consumers. Educators can promote sustainable practices by introducing it in the school curriculum, encouraging schools to implement food programs that emphasize fresh, healthy, and locally sourced food, and encouraging students to engage in community gardens and other food system initiatives. They can encourage students to learn about environmental and social challenges facing the modern food system and ways to address them. Educated consumers can take informed decisions while supporting sustainable practices and rejecting unethical ones.

Consumers play a crucial role in promoting sustainable and equitable food systems. The modern food system is supply-driven, with corporations producing food that is often harmful to the environment and the consumer's health. Consumers must demand transparency from producers, including information about the origin, nutritional content, and production method of food products. Labeling laws must be enacted to enable consumers to make informed decisions about the food they eat. Consumers should be encouraged to buy locally sourced, in-season produce, which attenuates the environmental impact of food transport and supports local agriculture. They should also prioritize buying from producers who prioritize sustainability and ethical labor practices. Consumers should be educated on food waste and encouraged to minimize it. Consumers must be encouraged to support grassroots initiatives that promote sustainable food behaviors, such as community-supported agriculture, food co-ops, and farmers markets. Promoting sustainable and equitable food systems require a multifaceted approach that involves policymakers, educators, and consumers. To promote sustainability, policymakers must develop legislation, incentives, and technical assistance programs that support sustainable agricultural practices. To promote equity, they must also enact laws that promote fair labor practices and increase access to healthy food. Educators must include sustainable food practices in the curriculum and promote community engagement, ensuring that consumers are exposed to sustainable food practices early on. Consumers must demand transparency from producers and make informed decisions regarding the food they eat. With the collective efforts of all stakeholders, sustainable and equitable food systems can be achieved, mitigating the environmental and social challenges within the modern food system.

A FINAL THOUGHT ON THE POTENTIAL OF LAB-GROWN MEAT AND VERTICAL FARMING TO CHANGE THE WAY WE EAT

The potential of lab-grown meat and vertical farming to change the way we eat is immense. These innovations offer a solution to the current unsustainable food production system and pave the way for a more sustainable and ethical food industry. With vertical farms, we can grow crops in a controlled environment, using less land, water, and pesticides, and reducing the carbon footprint of agriculture. This method of food production can also guarantee year-round crop yield and reduce the dependence on fossil fuels used in farming machinery. Meanwhile, lab-grown meat eliminates the need to raise and slaughter animals while reducing the environmental impact of meat production. This technique requires only a fraction of the resources required to produce the same amount of meat from traditional animal farming, thus reducing greenhouse gas emissions and pollution associated with animal agriculture. The benefits of lab-grown meat and vertical farming extend beyond environmental sustainability. These innovations also offer a solution to some of the ethical concerns regarding food production. By removing the need for animal farming and reducing the need for animal testing in the food industry, we can significantly reduce animal suffering. The lab-grown meat industry offers a solution to the increasing demand for meat consumption while reducing the ethical dilemmas associated with meat production. There are still some challenges that need to be addressed before we can embrace these innovations fully. Firstly, the current production costs of lab-grown meat make it more expensive than traditional meat production methods. The cost of production needs to be lower for lab-grown meat to be a viable substitute for traditional meat. Also, there are health concerns regarding the use of artificial growth hormones and the safety of cellular agriculture. These factors must be addressed before lab-grown meat can be accepted as a legitimate food source. Vertical farming requires a substantial initial investment, and the technology might not be accessible to small farmers or low-income communities, leading to issues of food inequality. It is not clear if these innovations will completely replace or coexist with traditional farming methods. While vertical farming has the potential to increase crop yield and reduce land use, traditional farming has been around for centuries and is deeply ingrained in society. Similarly, while lab-grown meat has numerous benefits, some people might not be willing to switch to synthetic meat, and it may take time for it to replace traditional meat entirely. There is no doubt that the future of food is changing through innovative technologies like lab-grown meat and vertical farming. These innovations have the potential to revolutionize the way we produce and consume food, addressing environmental, ethical, and health challenges associated with traditional food systems. While there are financial, technological, and societal challenges that need to be addressed, it is crucial to recognize the potential of these innovations and support further research and development in this field. Ensuring the accessibility and affordability of these innovations can go a long way in advancing food sustainability and promoting social and environmental justice. The integration of both innovation and old school methods may lead to better food quality, affordable prices, and sustainability, all while still staying true to our cultural roots. The future of food production is rapidly evolving, with new technologies and innovations emerging every day. Two major developments in this field include lab-grown meat and vertical farming, both of which offer promising solutions to many of the challenges that face our traditional agricultural system. While these two approaches are often seen as separate, they share a common goal – to produce food more efficiently, sustainably, and safely. Lab-grown meat, or cultured meat, involves growing meat cells in a lab, rather than raising animals for slaughter. This is achieved by first obtaining muscle cells from an animal biopsy, usually from a living donor. These cells are then placed in a culture medium that contains the necessary nutrients and growth factors to allow them to proliferate and differentiate into muscle tissue. The resulting product is nutritionally identical to conventionally produced meat but is produced in a controlled environment that offers many benefits over traditional farming methods. One of the primary advantages of lab-grown meat is its potential to reduce the environmental impact of the meat industry. Conventional livestock production is notoriously resource-intensive, requiring vast amounts of land, water, and feed to produce the same amount of protein as cultured meat. Livestock production is a major contributor to greenhouse gas emissions and other environmental problems, such as deforestation and water pollution. By contrast, lab-grown meat production can be much more efficient, requiring less land and water, and generating far fewer emissions and waste products. Another benefit of lab-grown meat is its potential to improve animal welfare. The conditions of factory farms are notoriously poor, and many people object to the suffering and exploitation of animals that takes place in these facilities. While cultured meat still requires the use of animal cells, it eliminates the need for the large-scale confinement, transportation, and slaughter of livestock that is so problematic. It allows for the production of meat from rare or endangered species, without causing harm to these animals or disrupting their ecosystems. Despite these benefits, lab-grown meat is still in its infancy and there remain many challenges to be addressed before it can become a viable alternative to traditional farming. One of the major obstacles is the cost of production, which is currently prohibitively high compared to conventional meat production. The technology and infrastructure required to produce cultured meat at scale are still under development, and regulatory and cultural barriers must also be overcome before it can gain widespread acceptance.

Vertical farming is another innovation in food production that offers many potential benefits. This approach is based on the idea of growing crops in vertically stacked layers, often in urban or indoor environments where space is at a premium. By utilizing high-tech systems such as hydroponics, aeroponics, and LED lighting, vertical farms can produce large amounts of food in a small footprint, with minimal resource inputs and environmental impact. One of the primary advantages of vertical farming is its potential to increase food security and accessibility. With global populations rising and arable land becoming scarcer, there is an urgent need to find more efficient ways of producing food. By utilizing vertical space, cities can produce fresh, healthy food locally, reducing dependence on food imports and improving food access for low-income communities. Vertical farms can grow crops year-round, in any climate, reducing the risk of crop failure due to weather events and climate change. Another benefit of vertical farming is its potential to reduce some of the environmental impacts associated with conventional agriculture. Traditional farming practices are often associated with soil erosion, water depletion, and pesticide and fertilizer use, which can lead to pollution and ecological damage. By contrast, vertical farming systems can utilize closed-loop systems that recycle water and nutrients, eliminate the need for soil, and greatly reduce pesticide use. Vertical farms can be located in or near urban centers, reducing transport costs and emissions associated with long-distance food transportation. Like lab-grown meat, vertical farming still faces significant challenges and limitations. One major obstacle is the high capital cost of setting up a vertical farm, which requires significant investments in technology and infrastructure. Energy costs can be high, as LED lighting and climate control systems are required to maintain optimal growing conditions. There are also limitations on the types of crops that can be grown using hydroponic or aeroponic systems, which may be unsuitable for certain crops or soil-based agriculture.

Despite these challenges, both lab-grown meat and vertical farming offer exciting potential for the future of food production. These innovations represent a shift away from traditional methods of farming and animal husbandry, and towards more sustainable and efficient systems that prioritize environmental, social, and ethical concerns. While there are still many obstacles to overcome, it is clear that the future of food will be shaped by these and other pioneering technologies that seek to create a more resilient and equitable food system for all.

BIBLIOGRAPHY

Paleontographical Society (Great Britain). 'Annual Report of the Paleontographical Society, ..., with List of the Council, Secretaries, and Members and a List of the Contents of the Volumes Already Published.' Paleontographical Society, 1/1/1896

Madsen, Ana Oliveira. 'Anthropological Approaches to Understanding Consumption Patterns and Consumer Behavior.' Chkoniya, Valentina, IGI Global, 4/3/2020

Board on Health Care Services. 'Health Professions Education.' A Bridge to Quality, Institute of Medicine, National Academies Press, 7/1/2003

Sue Booth. 'Critical Dietetics and Critical Nutrition Studies.' John Coveney, Springer, 2/13/2019

Health and Medicine Division. 'Building a More Sustainable, Resilient, Equitable, and Nourishing Food System.' Proceedings of a Workshop, National Academies of Sciences, Engineering, and Medi- cine, National Academies Press, 7/3/2021

Nathan A. Rosenberg. 'Farming for Our Future.' The Science, Law, and Policy of Climate-neutral Agriculture, Peter H. Lehner, Environmental Law Institute, 1/1/2021

Julian Roche. 'Agribusiness.' An International Perspective, Routledge, 8/6/2019

Steve Wells. 'A Very Human Future.' Enriching Humanity in a Digitized World, Rohit Talwar, Fast Future Publishing Ltd , 10/15/2018

Dr. Dickson Despommier. 'The Vertical Farm.' Feeding the World in the 21st Century, Macmillan, 10/12/2010

Rajeev Bhat. 'Future Foods.' Global Trends, Opportunities, and Sustainability Challenges, Academic Press, 12/4/2021

Ruth Feber. 'Wildlife Conservation on Farmland: Managing for nature on lowland farms.' David Whyte Macdonald, Oxford University Press, 1/1/2015

Andrew Hessel. 'The Genesis Machine.' Our Quest to Rewrite Life in the Age of Synthetic Biology, Amy Webb, Public Affairs, 2/15/2022

John Robbins. 'The Food Revolution.' How Your Diet Can Help Save Your Life and Our World, Mango Media Inc., 9/15/2010

Benjamin Aldes Wurgaft. 'Meat Planet.' Artificial Flesh and the Future of Food, Univ of California Press, 10/13/2020

Paul Shapiro. 'Clean Meat.' How Growing Meat Without Animals Will Revolutionize Dinner and the World, Simon and Schuster, 1/2/2018

Mohamed Merzouki. 'Nutrition and Human Health.' Effects and Environmental Impacts, Hicham Chatoui, Springer Nature, 6/28/2022

George Watson. 'Writing a Thesis.' A Guide to Long Essays and Dissertations, Longman, 1/1/1987

Peter Oosterveer. 'Global Governance of Food Production and Consumption.' Issues and Challenges, Elgar, 1/1/2007

www.ingramcontent.com/pod-product-compliance
Lightning Source LLC
Chambersburg PA
CBHW072359290526
45794CB00001B/115